T0291838

CAMBRIDGE LIBRARY COLLECTION

Books of enduring scholarly value

Botany and Horticulture

Until the nineteenth century, the investigation of natural phenomena, plants and animals was considered either the preserve of elite scholars or a pastime for the leisured upper classes. As increasing academic rigour and systematisation was brought to the study of 'natural history', its sub-disciplines were adopted into university curricula, and learned societies (such as the Royal Horticultural Society, founded in 1804) were established to support research in these areas. A related development was strong enthusiasm for exotic garden plants, which resulted in plant collecting expeditions to every corner of the globe, sometimes with tragic consequences. This series includes accounts of some of those expeditions, detailed reference works on the flora of different regions, and practical advice for amateur and professional gardeners.

The Forest Trees of Britain

A keen collector and sketcher of plant specimens from an early age, the author, educator and clergyman Charles Alexander Johns (1811–74) gained recognition for his popular books on British plants, trees, birds and countryside walks. *Flowers of the Field* (1851), one of several works originally published by the Society for Promoting Christian Knowledge, is also reissued in this series. First published by the Society between 1847 and 1849, this two-volume botanical guide for amateur enthusiasts focuses on the trees found in British woodland. Johns describes each species, noting also pests and diseases, uses for the wood, and associated myths and legends. The work is noteworthy for its meticulous engravings of leaves, seeds and blossom, and of the trees in natural settings. Volume 2 (1849) covers such species as chestnut, birch, elm, willow, ivy, yew and cedar. An index to the whole work is also provided.

Cambridge University Press has long been a pioneer in the reissuing of out-of-print titles from its own backlist, producing digital reprints of books that are still sought after by scholars and students but could not be reprinted economically using traditional technology. The Cambridge Library Collection extends this activity to a wider range of books which are still of importance to researchers and professionals, either for the source material they contain, or as landmarks in the history of their academic discipline.

Drawing from the world-renowned collections in the Cambridge University Library and other partner libraries, and guided by the advice of experts in each subject area, Cambridge University Press is using state-of-the-art scanning machines in its own Printing House to capture the content of each book selected for inclusion. The files are processed to give a consistently clear, crisp image, and the books finished to the high quality standard for which the Press is recognised around the world. The latest print-on-demand technology ensures that the books will remain available indefinitely, and that orders for single or multiple copies can quickly be supplied.

The Cambridge Library Collection brings back to life books of enduring scholarly value (including out-of-copyright works originally issued by other publishers) across a wide range of disciplines in the humanities and social sciences and in science and technology.

The Forest Trees of Britain

VOLUME 2

CHARLES ALEXANDER JOHNS

CAMBRIDGE
UNIVERSITY PRESS

CAMBRIDGE
UNIVERSITY PRESS

University Printing House, Cambridge, CB2 8BS, United Kingdom

Published in the United States of America by Cambridge University Press, New York

Cambridge University Press is part of the University of Cambridge.
It furthers the University's mission by disseminating knowledge in the pursuit of
education, learning and research at the highest international levels of excellence.

www.cambridge.org
Information on this title: www.cambridge.org/9781108069151

© in this compilation Cambridge University Press 2014

This edition first published 1849
This digitally printed version 2014

ISBN 978-1-108-06915-1 Paperback

This book reproduces the text of the original edition. The content and language reflect
the beliefs, practices and terminology of their time, and have not been updated.

Cambridge University Press wishes to make clear that the book, unless originally published
by Cambridge, is not being republished by, in association or collaboration with, or
with the endorsement or approval of, the original publisher or its successors in title.

THE

FOREST TREES OF BRITAIN.

BY THE

REV. C. A. JOHNS, B.A., F.L.S.,

AUTHOR OF

"A WEEK AT THE LIZARD," "BOTANICAL RAMBLES," ETC.

PUBLISHED UNDER THE DIRECTION OF
THE COMMITTEE OF GENERAL LITERATURE AND EDUCATION,
APPOINTED BY THE SOCIETY FOR PROMOTING
CHRISTIAN KNOWLEDGE.

IN TWO VOLUMES.

VOLUME II.

LONDON:

PRINTED FOR

THE SOCIETY FOR PROMOTING CHRISTIAN KNOWLEDGE,

SOLD AT THE DEPOSITORY,

GREAT QUEEN-STREET, LINCOLN'S INN-FIELDS, AND 4, ROYAL EXCHANGE;

AND BY ALL BOOKSELLERS.

1849.

CONTENTS.

LIST OF ILLUSTRATIONS.

viii   LIST OF ILLUSTRATIONS.

THE CHESTNUT.

THE

FOREST TREES OF BRITAIN.

THE CHESTNUT.

CASTANEA VESCA.

Natural order—AMENTACEÆ.

Class—MONŒCIA. *Order*—POLYANDRIA.

BEFORE I begin the description of this the
most magnificent tree which reaches perfection in
Europe, it is necessary that I should examine somewhat minutely the grounds which have been
urged in favour of its claims to be considered a
native of Great Britain, which are neither few nor
inconsiderable.

The first of these is derived from the large
quantity of Chestnut timber which, it has been
said, exists in old buildings. Evelyn, writing on
this subject, says: "The Chestnut is, next the
Oak, one of the most sought after by the car-
penter and joiner. It hath formerly built a good
part of our ancient houses in the city of London,
as does yet appear: I had once a very large barn
near the city, framed entirely of this timber; and
certainly the trees grew not far off, probably in
some woods near the town; for in that description
of London, written by Fitz-Stephen, in the reign

of Henry II., he speaks of a very noble and large
forest which grew on the boreal part of it; ' Nigh to
London,' says he, ' extends a huge forest, the woody
resort of wild beasts, a hiding place for deer, boars
and wild bulls, &c.' * A very goodly thing it seems,
and as well stored with all sorts of good timber as
with venison and all kinds of chase; and yet some
will not allow the Chestnut to be a free-born of
this island, but of that I make little doubt."

Dr. Ducarel, in his Anglo-Norman Antiquities
observes, that " many of the old houses in Normandy when pulled down are found to have a
great deal of Chestnut timber about them. As
there are not any forests of Chestnut-trees in
Normandy, the inhabitants have a tradition that
this timber was brought from England; and there
are some circumstances, which, when rightly considered will add strength to this tradition; for
many of the old houses in England are found to
contain a great deal of this timber; several of the
houses in Old Palace Yard, Westminster, and in
that neighbourhood, which were taken down in
order to build Parliament and Bridge Streets,
appeared to have been built with Chestnut."

Hasted, who contributed to the Philosophical
Transactions a letter confirmatory of Ducarel's
views, says: " The ancient Norman buildings are
mostly of this wood, which in all probability was
fetched home from this country; most of the
stone wherewith our monasteries and buildings
of such sort were erected came from Normandy.
This seems to have been a mutual traffick for
some centuries between the two countries."

* Proxime patet foresta ingens, saltus nemorosi ferarum, latebræ
cervorum, damarum, aprorum, et taurorum sylvestrium.

Sir Thomas Dick Lauder mentions, that " the roof of the Parliament House in Edinburgh is constructed of it, and the beams, and roofing, and strange projections of many of the wooden houses, which had stood for ages in the ancient part of the Scottish capital, and which were recently pulled down, were found to be of Chestnut; and what is curious, the timber seems to have been procured from a suburban forest, resembling that on the north side of ancient London; for it appears, from the city records, that large Oaks and Chestnuts formerly covered the space called the Borough-moor, a wild piece of ground, then lying about two miles to the south-west of the city." Gilpin also states, that he had seen in the belfry of the church at Sutton, near Mitcham in Surrey, beams like Oak, yet plainly appearing to be of a different kind of timber, and supposed to be Chestnut.

Another argument in favour of this opinion is derived from the fact that there are in England several places which take their name from these trees, consequently that the trees must have grown there in considerable abundance before such names were given. Such are, Norwood Chesteney in the parish of Milton near Sittingbourne, and Chestnut Hill near the same place. In Hertfordshire is a town called in old writings Cheston, Chesthunte, Shesterhunte and Cestrehunt; and Philpot, who wrote in 1659, says: " There is a manor called Northwood Chasteners, which name complies with the situation, for it stands in a wood where Chestnut-trees formerly grew in abundance."

Evidence still more direct is afforded by the mention of trees existing in a living state, at

periods of time more or less remote. For example, there was in the parish of Milton in Elizabeth's time, a Chestnut-wood containing 278 acres,
and called Cheston. The forest near London described by Fitz-Stephen is also quoted by Miller,
Lauder and others, but unfortunately, as we shall
see by and by, without examining the original
author. The great Tortworth Chestnut in Gloucestershire, to be described hereafter, was known by
that name as early as the reign of Stephen; and in
the confirmation of a grant made by Henry II., to
the monks of Flexeley, " the tithe of Chestnuts in
the Forest of Dean" is secured to the monastery.

Such are the principal arguments in favour of
the opinion that the Chestnut is a native tree. On
the other side it is urged, that the name Spanish
Chestnut would imply that the tree is of foreign
origin. But this argument will not bear examination. There can be no doubt that Chestnuts
were imported from Spain at the time when the
name was given; but it does not at all follow that
none were produced in England: we are equally
justified in inferring that they were so called by
merchants, to distinguish them, and recommend
them above English Chestnuts, which are far inferior, just as we call hazel-nuts imported from
the same country, " Spanish nuts," to distinguish
them from those which grow in England.

In the next place, it is expressly stated by ancient authors that the Chestnut-tree was first introduced from Asia into Europe by the Greeks,
and transported thence into Italy by the Romans.
This fact is allowed by the holders of the opposite
opinion, who at the same time maintain that the
tree might be unknown in continental Europe,

and yet be indigenous to Britain, although un-
noticed by the invaders of that country.

The fact that Chestnut timber has been found
in ancient buildings in very great quantities would
carry great weight, but that it has been recently
discovered that the wood supposed to be Chestnut
is in reality a kind of Oak, differing from common
Oak in those very characters which had been
fixed on as distinctive of Chestnut. Besides this,
Chestnut timber of large dimensions, is, neither
in Great Britain nor the south of Europe, found
to possess the qualities, strength and durability,
which were supposed to have recommended it to
the notice of ancient builders.

Evelyn's quotation from Fitz-Stephen is a very
unhappy one, and the citation of the same passage
from Evelyn by Miller, Laud, &c., still more unfortunate, for the tree in question is neither described nor even mentioned by name. Evelyn
honestly cited the passage as evidence that there
formerly existed a great forest near London, in
which he thought it probable that Chestnut timber,
among other kinds, might grow, and the authors
who followed him, perhaps from not being able
to refer to the original work, mistaking the drift
of his remark, took it for granted that the tree
was mentioned, and considered the evidence there-
by afforded conclusive, as indeed they well might.

There can be no doubt that Chestnut-trees have
existed quite long enough in England to originate
the names of places, but this they might have done
without being aboriginal trees. In fact a planted
grove of foreign trees, or even a single fine specimen, might have afforded sufficient reason for giving a name commemorative of the circumstance to

an otherwise unimportant place. With respect to
the Chestnut-forest said to exist in Elizabeth's time
in the parish of Milton, Barrington, who wrote in
1771, says, that he expended much time and labour
in examining the forest, and discovered satisfac-
tory evidence, from the fact that the trees stood at
equal distances from each other and in straight lines,
that the trees had been originally planted. The
author of a tract published in 1612 was evidently
of opinion that the tree in question was not indigenous, for he recommends planting it as a " kind
of timber tree of which few grow in England."

With regard to the strongest evidence of all in
favour of the opinion that the Chestnut is a native
tree, that, namely, afforded by the actual existence
of ancient trees, and the notice of others in the
grant to the Monastery of Flexely; * it may be
argued : that, supposing the Chestnut to have
been introduced by the Romans, ample time had
been allowed it to establish itself thoroughly, and
even to spread itself over the country. The Sycamore, Gerard says, was, in 1597, a " rare exotic,"
yet 250 years have sufficed so thoroughly to natu-
ralize it, that few persons are aware that it is
really of foreign origin. Three times that space
of time may have elapsed between the introduction of the Chestnut and the first mention of a
British specimen, so that even if Fitz-Stephen had
told us that the forest near London consisted of
these trees, it would not necessarily follow that
they were not descended from trees originally introduced. That they did not exist in great quantities may I think be inferred from the fact that

* " Singulis annis totam decimam Castanearum de Denâ, et terram
illam quam adquietavit ipse Comes Herefordiæ."

the produce, with the exception of a tithe, was considered so important as to be reserved by the king. Had the tree been, as Evelyn surmises, abundant near London, the forest of Dean would scarcely have been laid under contribution; the fact, therefore, that Chestnuts are mentioned at all would afford evidence rather that they were rare and consequently valuable, than that they were common forest trees.

On the whole then we may, I think, with reason conclude that the chestnut, though long naturalized in England, is not an aboriginal native, but was introduced probably by the Romans at a very early period, and in process of time propagated itself so widely as to have raised a doubt whether it was not a really native tree. Its history may be briefly told as follows. It was first introduced into Europe by the Greeks from Sardis in Asia Minor, whence it was called the "Sardian nut,"* and at a later period, "Jupiter's nut,"† and "husked nut" from its being enclosed in a husk or rind instead of a shell. Loudon, and several other authorities, from a misconception of a passage in Pliny, or, more likely, from quoting it at second hand, attribute the introduction of this tree into Italy to Tiberius Cæsar, a gross inaccuracy, for it is evident from the writings of Virgil that chestnuts were abundant in Italy long before the time of that emperor. By the Romans it was called Castanea, from Castanum, a town of Magnesia, in Thessaly, where it grew in great abundance, and from which it is said that they first brought it. Pliny enumerates several varieties, the best of which he says grew at Tarentum and

* *Sardianus balanus.* PLIN. † Διὸς βαλάνος. THEOPH.

Naples. Theophrastus, who wrote in the third century before the Christian era, speaks of it under the name of Jupiter's nut, as a tree originally introduced, but in his time quite naturalized in the mountainous parts of Thessaly.

From Italy and Greece it appears to have spread over the greater part of temperate Europe, ripening its fruit and sowing itself wherever the grape ripens. It was in all probability introduced into Britain by the Romans for the sake of its fruit; and here, from being a tree of great duration, and from the paucity of other trees the fruit of which is available for food, it was naturally an object of care and attention. In France, Italy, and Spain, especially the two last countries, it attains a great size, and has all the appearance of being naturalized. On the Alps and Pyrenees it flourishes at an elevation of between 2,500 to 2,800 feet, the nuts having been perhaps carried to these lofty situations by the animals which lay up stores of winter food. It is still most abundant in Asia Minor, as well as in Armenia and Caucasus, and it is also found in America, as far north as latitude 44°. It ripens its fruit in the warmer parts of Scotland; but rarely, if at all, in Ireland.

The Chestnut-tree is twice mentioned in the Authorized Version of the Old Testament (Gen. xxx. 37, and Ezek. xxi. 8): but in the former of these passages the Septuagint translation renders the Hebrew word *armon*, by *plane*, in the latter by *pine*. Rosenmuller is of opinion that the rendering "plane" is the correct one.

The Chestnut was, by Linnæus, placed in the same genus with the Beech, under the name of *Fagus castanea*, but modern botanists have again

separated them, considering the genera sufficiently
distinguished by the former having the barren
flowers on long spikes, and producing farinaceous
nuts; the latter, by having its barren flowers in
globular heads, and by bearing oily nuts—characters strong enough to mark different genera.
It is well distinguished by its large, sharply serrated leaves, which are smooth and glossy, by
its long tendril-like spikes of flowers in July
and in autumn by its bunches of nuts enclosed
in cases thickly beset with complicated sharp
prickles. Sir James E. Smith describes it as a
stately and majestic tree, rivalling, if not exceeding the British Oak in size and duration. The
bark is remarkable for its deep and wide clefts,
which seem to have furnished ideas for some
ornaments in Gothic Architecture. Gilpin, who
also compares it to the Oak, says that its ramification is more straggling but easy, and its foliage
loose. This is the tree which graces the landscapes of Salvator Rosa. In the mountains of
Calabria, where Salvator painted, the Chestnut
flourished. There he studied it in all its forms,
breaking and disposing it in a thousand beautiful
shapes, as the exigences of his composition required. We find it indeed nearly always forming
a prominent feature in his bold and rugged landscapes, many of his most striking scenes being
drawn from the wild haunts and natural fastnesses
of that romantic country, wherein he passed so
many of his youthful days. Gilpin supposes that
this great painter's fondness for the Chestnut is
owing to its liability to be shattered by storms.
Bosc is of opinion, that as an ornamental tree it
ought to be placed before the Oak. Its beautiful

leaves, he says, which are never attacked by insects, and which hang on the trees till very late
in autumn, mass better than those of the Oak and
give more shade. An old Chestnut, standing alone,
produces a superb effect. A group of young
Chestnuts forms an excellent back ground to
other trees; but a Chestnut-coppice is insupport-
ably monotonous.

But it is in Italy that it is to be seen in all
its grandeur. Sir T. D. Lauder speaks of having
"roamed for miles through the high-roofed leafy
shades of the endless Chestnut-forests, which hung
everywhere on the sides and roots of the Apen-
nines, where the impervious canopy was supported
by the columnar trunks of the enormous trees;
and there, and in many parts of the Alps, the
peasants depend greatly on the Chestnuts; for
the bread they live on is very much, if not altogether, composed of the farina obtained from the
nuts. We remember participating in one of the
most interesting scenes we ever beheld, whilst penetrating that extensive Chestnut-forest which
covers the body of the Valombrosan Apennine,
for nearly five miles upwards. It was a holiday,
and a group of peasants, of both sexes, dressed in
the gay and picturesque attire of the neighbourhood of the Arno, were sporting and dancing on a
piece of naturally level and well cropped turf,
which spread itself beneath these gigantic trees,
whilst the inmost recesses of the forest were, ever
and anon, made to resound to their mirth and
their music. Some were beating down the Chestnuts with sticks; others, for their own refresh-
ment, were picking out the contents from the
pallisadoed castles in which the kernels lie in-

trenched; and when newly gathered from the tree, nothing can be more sweet or pleasing to the palate: whilst others, and particularly the girls, were carrying on an amusing warfare of love, by pelting one another with the fruit. It seemed to us as if the golden age had been restored; and that, abandoning all the luxuries and attendant evils of civilized life, mankind had voluntarily returned to their pristine simplicity of fare, when the *esculus* and the Chestnut-tree yielded them their innocuous food, and when the innocency of their lives corresponded with that of their rustic nutriment."

The Chestnut will thrive in most situations, except where the soil is stiff and tenacious; it prefers a deep sandy loam, but, as we have seen, attains a great size, at a considerable elevation among the mountains of the south of Europe. In England, it grows with the greatest rapidity in the rich loamy soils of the valleys, but its timber is then brittle and useless; in sheltered situations, where the soil is tolerably free, it attains some height, but in poor gravelly soil, where its roots will only run along the surface, the trunk attains a considerable diameter, with a disproportionate spread of branches. Bosc remarks, that wherever he has observed Chestnuts on mountains in France, Switzerland, and Italy, they were never in soils or on surfaces fit for the production of corn; where the corn left off, there the Chestnuts began; and, in climates suitable for corn, the tree is only found on rocky and flinty soils.

According to Phillips, " the Chestnut seems to delight in the cinerated substances thrown out of

volcanoes, as is shown by the thick woods of
Chestnut-trees, which cover the surface in the
neighbourhood of Vesuvius. They grow luxuriantly on Mount Somma, on the heights of the
Camaldoli near Naples, on the Pyrenees near
medicinal springs, and in general, in the neigh-

FLOWER AND RIPE FRUIT OF CHESTNUT.

bourhood of subterraneous fires, not to mention
the numerous and gigantic trees that have for
ages darkened the sides of Etna."

The Chestnut is usually propagated by sowing the nuts; but the choice varieties are continued by grafting. English nuts germinate freely, but if foreign ones be sown, care should be taken to ascertain that they have not been kiln-dried, as in that case the germinating principle is destroyed.

The Chestnut blossoms in July, and soon the upper part of the spike bearing the barren flowers withers, and drops off, leaving the lower part of the spike still supporting the fertile flowers, with the embryo of the future nuts attached. Towards

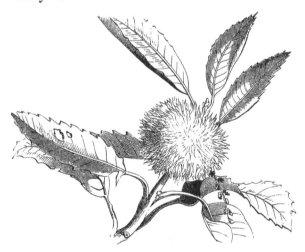

FRUIT OF CHESTNUT.

the end of September the latter begin to ripen, and in October fall to the ground, where they open with valves and expose the ripe nuts. Each case contains from two to five nuts, two or more of which are often mere empty rinds, but all, whether solid or otherwise, bear the remains of the flower, in the shape of a few dry bristles, on their points.

Chestnuts are spoken of very contemptuously by Pliny, who says: "The fruit of the Castanea we call also a nut, though it approaches nearer in character to mast. It is protected by a case beset with strong prickles. It is strange that we hold as of no value a fruit which Nature has so carefully guarded from injury. As many as three nuts frequently grow together in one case. The proper rind of the nut is tough, and within this is a thin skin closely attached to the substance of the nut, as in the walnut, which, unless it be removed, spoils the flavour of the fruit. The best way of preparing them for food is by roasting. They are sometimes ground into meal, which is converted by women into a wretched substitute for bread, and eaten during their religious fasts."

Notwithstanding the low repute in which the Roman naturalist appears to have held chestnuts, they are mentioned among rustic dainties by more than one Latin poet. Thus Virgil says:

> " ——Sunt nobis mitia poma,
> Castaneæ molles, et pressi copia lactis."

> " Ripe apples and soft chestnuts we have there,
> And curd abundant to supply our fare."

> " Ipse ego cana legam tenera lanugine mala,
> Castaneasque nuces, mea quas Amaryllis amabat,
> Addam cerea pruna."

> " Myself will search our planted grounds at home,
> For downy quinces and the waxen plum ;
> And pick the chestnuts in the neighbouring grove,
> Such as my Amaryllis used to love."

And Martial:

> " Et quas docta Neapolis creavit,
> Lento castaneæ vapore tostæ."

> " And chestnuts, such as learned Naples boasted,
> With simmering heat by careful housewife roasted."

Other authors speak of an improved variety produced by grafting the Chestnut on the Almond. The fruit thus produced, they say, has a less prickly rind. Thus Palladius:

> " Castaneamque trucem depulsis cogit echinis
> Mirari fructus lævia poma sui."

> " Bids the rough Chestnut change its prickly kind,
> And deck the tree with balls of polished rind."

A slight acquaintance with physiological botany will suffice to show the impossibility of this change, the two trees being too dissimilar to allow their being grafted on one another.

Our own poets make frequent mention of roasted chestnuts. Thus Ben Jonson speaks of the " chestnut whilk hath larded (fattened) many a swine;" and Shakspeare, of " the sailor's wife with chestnuts in her lap:" and Milton, writing on the death of his friend, Deodati, says:

> " In whom shall I confide ?   Whose counsel find
> A balmy medicine for my troubled mind ?
> Or whose discourse, with innocent delight
> Shall fill me now, and cheat the wintry night—
> While hisses on my hearth the purple pear,
> And blackening chestnuts start and crackle there ;
> While storms abroad the dreary meadows whelm,
> And the wind thunders through the neighbouring Elm ? "

Evelyn laments that in his time they were not used as an article of food so much as they deserved. " We give that fruit to our swine

II.   C

in England, which is amongst the delicacies of
princes in other countries, and, being of the larger
nut, is a lusty and masculine food for rusticks at
all times, and of better nourishment for husband-
men than cole (cabbage) and rusty bacon, yea, or
beans to boot; instead of which, they boil them
in Italy with their bacon; and in Virgil's time
they eat them with milk or cheese. The bread
of the flour is exceeding nutritive; it is a robust
food, and makes women well complexioned, as I
have read in a good author. They also make
fritters of chestnut-flour, which they wet with
rose-water, and sprinkle with grated parmigiano,
and so fry them in fresh butter for a delicate.
How we here use chestnuts in stewed meats and
beatille pies, our French cooks teach us; and
this is in truth their very best use, and very com-
mendable."

The principal countries where chestnuts are
now employed as an important article of food are
the south of France and the north of Italy; where
they serve, in great measure, as a substitute for
both the bread and potatoes of more northern nations. In these countries it becomes a matter of
importance to preserve the chestnuts during winter; and accordingly great care is taken in gathering, keeping, and drying them so as to ensure a
constant supply. When the chestnuts are ripe,
those that are to be preserved are collected every
day from the ground on which they have fallen
from the tree, and spread out in a dry airy place,
till the whole are gathered together. But as it is
often a considerable time before the chestnuts are
all ripe enough to fall from the tree, if the season
be so far advanced that there is danger of snow or

heavy rains, after the fallen chestnuts have been collected and set on one side for drying, the tree is beaten with long poles, to knock off the remaining fruit. But the fruit thus collected is only considered fit for immediate use; and the greater part is carried to the local markets or sent to Paris. The husks of the chestnuts beaten off the trees being generally attached to the nuts, they are trodden off by peasants furnished with heavy sabots, or wooden shoes, when the nuts are wanted for present use; but when they are to be preserved for a few months, they are generally kept in their husks in heaps in the open air, or in barrels of sand, which are sometimes actually sprinkled with water in very dry seasons, in order to preserve the full and plump character of the nuts.

In the Cevennes, where chestnuts are an article of food, the inhabitants have a process of kiln-drying them, so that they will keep good for two or three years. The process consists in exposing them on the floor of a kiln to the smoke of a smothered wood fire. The heat is applied gently, so as to make the internal moisture transpire through the husk of the chestnut. The fire is kept gentle for two or three days, and then is gradually increased during nine or ten days. The chestnuts are then turned with a shovel, and the fire is continued until they are ready. This is known by taking out a few of them and threshing them; if they quit their inner skin, they are done. They are then put into a bag, and threshed with sticks to separate the external and internal husks. If the husks are left on, the chestnuts become black, by imbibing from the husk the empyreumatic oil of the wood-smoke, and do not keep so well. In order to be prepared for food, they are ground into flour, and of this, mixed up with a little milk and salt, and sometimes with the addition of eggs and butter, is made a thick girdle-cake, called *la galette*. *La polenta* is another preparation made by boiling the chestnut-flour in milk till it becomes quite thick; when made with water, it is eaten with milk in the same manner as oatmeal porridge in the north of England and Scotland.

The most usual modes of cooking chestnuts in France are, boiling them in water, either simply, with a little salt, or with leaves of celery, sage, or any other herbs that may be preferred, to give them a flavour; and roasting them either in hot ashes, or in a coffee-roaster. They are also occasionally roasted before the fire, or on a shovel, as in England; but, when thus prepared, they are thought not so good. In whatever way they are roasted, the French cooks always slit the skin of all except one; and when that cracks and flies off, they know that the rest are done. Sugar is said to have been obtained in France from chestnuts, by the same process as is used for the extraction of sugar from beetroot, and at the rate of 14 per cent; which is more than the average produce of the beetroot. Chestnuts are sometimes used for whitening linen, and for making starch; and when roasted they are a good substitute for malt in making beer.

"In many countries," says Miller, "where Chestnut-trees are cultivated, the people graft the largest and fairest fruit upon stocks raised from the nut. And these grafted trees are called by the French *marroniers*, but they are unfit for timber." In France great attention is still paid to the cultivation of the Chestnut, and the varieties are divided into two sections, *les chataignes*, and *les marrons*, the latter being held in high esteem as producing nuts of the largest size, finest flavour, and farinaceous qualities.

The Chestnut-tree retains its leaves until late in the autumn, when they become of a rich golden hue. Owing to the tufted, and consequently weighty, character of the foliage, and the brittleness of the timber, the tree is liable to be injured by autumnal storms; but the leaves are rarely attacked by insects, strongly contrasting in this respect with its rival the Oak. Its leaves are in France used as litter for cattle, and, like those of the Beech, are sometimes employed for stuffing mattresses.

It is not easy to form a correct estimate of the value of the Chestnut as a timber-tree, there being a great uncertainty whether the beams discovered in ancient buildings, and said to be of Chestnut, do in every instance (they certainly do not in most cases) belong to this tree at all. On the whole, its value appears to have been much overrated, for, omitting the uncertain evidence afforded by ancient specimens, recent timber possesses few valuable properties, at least for the purposes for which it was recommended by the earlier writers. They, finding what they believed to be Chestnut-timber occurring in buildings of unquestionable antiquity, naturally concluded that its value for strength and durability recommended it to the earlier builders, and further assumed, that, owing to the estimation in which it was held, it had become rare. Thus Hartlib, who wrote before the middle of the seventeenth century, says:

" In divers places in Kent, as in and about Graves-
end, in the country and elsewhere, many prime
timbers of their old barns and houses are of Chestnut-wood, and yet there is now scarce a Chestnut-
tree within twenty miles of the place and the people altogether ignorant of such trees." And Evelyn, falling into the same error with regard to the
timber of which one of his barns was made, assumes that Chestnut-forests formerly stood in the
vicinity of London, and quotes as confirmation of
his surmise the passage from Fitz-Stephen cited
above, though that author makes no allusion to
the tree.

The French naturalist, Buffon, was the first
who directed attention to the strong resemblance
borne by the timber of the Durmast Oak (*Quercus sessiliflora*) to that of the Chestnut in its best
condition, both almost entirely wanting the silver
plates which characterise the timber of the common Oak (*Quercus pedunculata*). The truth of
this remark was subsequently confirmed by the
discoveries of Fougeroux and Daubenton in 1780,
and it is now an ascertained fact that the roof of
Wesminster Hall and other ancient buildings, formerly supposed to consist of Chestnut, is constructed of Durmast, or, as Lindley would have
it called, English, Oak. The fact is now ascertained to be, that Chestnut-timber, though admirably adapted in its young state for many purposes
to which Oak is applied, such as beams, posts, and
fences, after a certain, and that comparatively
an early, age becomes what is technically called
shaky and so deteriorated by the separation and
decay of the internal layers as to be of very little
value. It is evident therefore that before it could

have attained the size necessary for it to be employed in the structure of such buildings as Westminster Hall, it must have lost all the properties
which recommended it. If cut when not more
than fifty years old, it consists almost exclusively
of heart-wood, with a layer of alburnum, or sap-
wood, equalling in thickness the breadth of the
bark; but when suffered to stand beyond its full
growth it is, on good authority, the worst of all
timber, being more brittle and more apt to crack
and fly into splinters, than any other. In the
hop counties the growth of Chestnut-coppice is
much encouraged, poles from this tree and the
Ash being preferred to all others. Casks, it is
said, made of Chestnut-wood contribute much to
the colour and quality of the wine, as well as to
the preservation of it; the fermentation is slow,
and the wine made in those vessels is sweeter. It
has also the property of lasting a long time,
when used for water-pipes or other purposes under-ground. It is said also to be noxious to
spiders and other insects, but this virtue belongs
rather to the situation of Oak-beams in old buildings than to any quality actually residing in Chestnut. As fuel, it is not held in great estimation, and
the charcoal made from it, though not of first-rate
quality, is in some places greatly sought after for
forges. The bark, especially of young trees, is used
for tanning, and sells at half the price of Oak-bark.

Among the finest specimens of Chestnut-trees
now existing in this country, the Tortworth Chestnut occupies the first place, and according to all
accounts is the largest and oldest tree of any kind
growing in Great Britain. This fine and most
interesting relic of feudal times, tradition tells us,

THE TORTWORTH CHESTNUT.

was known as a boundary mark to the manor
of Tortworth, in Gloucestershire, as early as the
reign of King John, and was even called the
great Chestnut of Tortworth in the reign of Stephen. It had then probably been planted not
less than three hundred years, being consequently,
now at least a thousand years old. It was formerly much confined by the walls of the garden
within which it stood, but this incumbrance was
removed by Lord Ducie, who at the same time
applied fresh earth to its roots, which renewed its
vigour. Strutt gives a beautiful etching of it in
his " Sylva Britannica," of which the accompanying wood-cut is a copy. As he studied the tree with
a painter's eye, the reader will probably be glad
to have his description of it, which is as follows :
" Since the commencement of its existence, a
thousand years have rolled over its yet green
head. How is it possible, bearing this reflection
in mind, to look upon its gigantic trunk, and
widely-spreading arms, without feelings of reverence ! How many, not merely generations of
men, but whole nations, have been swept from
the face of the earth, whilst, winter after winter,
it has defied the howling blasts with its bare
branches; and, spring after spring, put forth its
leaves again, a grateful shelter from the summer
sun ! Its tranquil existence, unlike that of the
human race, stained by no guilt, chequered by
no vicissitudes, is thus perpetually renewing itself;
and, if we judge from the luxuriance of its foliage
and the vigour of the branches which encircle the
parent stem in wild profusion, may be prolonged
for as many more centuries as it has already stood.
Nor is it solitary in its old age. Its progeny

rises round it, and its venerable roots are nearly hidden by the lighter saplings and bushes that have sought the protection of its boughs, making it appear a grove in itself — a fit residence for some sylvan deity.

" It is only on approaching within the very limits overshadowed by its spreading branches, that the size of the majestic tree can be duly estimated; but when its full proportions are fairly viewed on all sides, it strikes the beholder with feelings of wonder and admiration, alike for its bulk and for the number of centuries which it has been in attaining it.

" It is only within a few years that it has been relieved from the pressure of three walls, in the angle of which it stood, a position which must have greatly injured the spreading of its roots. The axe, which might have been commendably employed in clearing the approach to it of brambles and briers, has, on the contrary, been barbarously, though not recently, applied to the tree itself, which has been wantonly despoiled of several large limbs on the north-east side, apparently many years ago; it is in consequence much decayed on that side, while on the other it is still sound. The Tortworth Chestnut, in 1766 measured fifty feet in circumference, at five feet from the ground. Its present measurement, at the same height, is fifty-two feet. The body is ten feet in height to the fork where it divides into three limbs, one of which, at the period already mentioned, measured twenty-eight feet and a half in girth, at the distance of five feet from the parent stem. The solid contents, according to the customary method of measuring timber, is

1965 feet; but its true geometrical contents must be much more. Young trees are now (1820) nursing from the nuts which it bore three years ago; and it is to be hoped that their pedigree will be preserved, as none can boast more ancient ancestry."

Loudon tells us, that in the year 1836 the tree was still in the same state as it was when drawn by Mr. Strutt.

Many other remarkable trees are described by different authors, but of these it will not be necessary to give more than a passing notice.

One mentioned by Gilpin grows at a place called Wimley, near Hitchin Priory, in Hertfordshire. In the year 1789, its girth, at five feet above the ground, was upwards of forty-two feet. Its trunk was hollow, and in part open; but its vegetation still vigorous.

The great Chestnut that stood at Finhaven, in Forfarshire, was long accounted the largest tree in Scotland. In the year 1760 a great part of the trunk of this remarkable tree, and some of its branches remained. In 1744 it measured, at six inches above the ground, nearly forty-three feet. "As this Chestnut," says Lauder, "appears from its dimensions, to have been planted about five hundred years ago; it may be presumed to be the oldest planted tree in Scotland."

Two very remarkable trees at Cobham in Kent are figured and described by Strutt in his "Sylva Britannica." One of these, called "The Four Sisters," from its four branching stems closely combined in one massive trunk, stands in the Heronry. It is the noble remains of a most magnificent tree; and though its head has paid for-

feit to the "skiey influences," during a long succession of revolving seasons, yet it is not left
entirely stripped of ornament in its old age; as a
number of tender shoots spring out of its topmost branches, and still give it, by the lightness
of their foliage, an appearance of freshness, of
which its aged trunk would almost forbid the expectation. It is thirty-three feet at twelve feet
from the ground, and forty feet at the point where
the trunk divides. It is not improbable that the
Four Sisters may have attained their tenth century. The other is not far off, and perhaps is
coeval with it. It is yet more of a wreck than
the first, one half of it lying, with shivered tops
and scattered boughs, stretched upon the ground.

At Cotehele, a seat of the Earl of Mount Edgecumbe, on the banks of the Tamar, there are
some exceedingly fine trees; one in particular,
though not equalling in dimensions those described above, is scarcely less imposing, from its not
showing external symptoms of decay. A few
feet above the ground it branches into three, each
of its giant limbs being the trunk of a lofty tree
covered with foliage to the summit.

"There was standing," says Evelyn, "an old
and decayed Chestnut at Fraiting, in Essex,
whose very stump did yield thirty sizeable loads
of logs. I could produce you another of the same
kind in Gloucestershire, which contains within
the bowels of it a pretty wainscoted room,
enlightened with windows and furnished with
seats."

The Chestnut-trees in Greenwich Park, some
of which are of great size, were planted by Evelyn,
and are therefore about two hundred years old.

A history of the Chestnut would be scarcely complete without a notice of the great tree near Mount Etna. It is called Castagno de Cento Cavalli, or, the Chestnut of a hundred horses from a traditionary tale that Joan of Arragon, when she visited Mount Etna, attended by her principal nobles, was obliged by a heavy shower to take refuge under this tree, the immense branches of which sheltered the whole party. Its fame had reached England before Evelyn's time, who quotes the following passage from Kircher: "And, what may seem scarce credible, the guide pointed out to me the shell of a single Chestnut-tree of such vast size, that a whole flock of sheep was enclosed in it by some shepherds, as in a very commodious fold." Brydone, who visited the spot in 1770, gives the following more detailed account: "From this place it is not less than five or six miles to the great Chestnut-trees, through forests growing out of the lava, in several places almost impassable. Of these trees there are many of an enormous size: but the Castagno de Cento Cavalli is by much the most celebrated. I have even found it marked in an old map of Sicily published near a hundred years ago; and in all the maps of Etna and its environs, it makes a very conspicuous figure. I own I was not much struck with its appearance, as it does not seem to be one tree, but a bush of five large trees growing together. We complained to the guides of the imposition; when they unanimously assured us, that by the universal tradition and unvarying testimony of the country, all these were once united in one stem; that their grandfathers remembered this, when it was looked upon as the glory of the forest and visited from all quarters; that for many years past it had been reduced to the venerable ruin we beheld. We began to examine it with more attention, and found that there is an appearance that these five trees were really once united in one. The opening in the middle is at present prodigious; and it does indeed require faith to believe that so vast a space was once occupied by solid timber. But there is no appearance of bark on the inside of any of the stumps, nor on the sides that are opposite to one another. Mr. Glover and I measured it separately, and brought it exactly to the same size, viz., *two hundred and four feet round.* If this was once united in one solid stem, it must with justice indeed have been looked upon as a very wonderful phenomenon in the vegetable world, and deservedly styled the glory of the forest.

" I have since been told by an ingenious ecclesiastic of this place that he was at the expense of carrying up peasants with tools to dig round the Castagno de Cento Cavalli, and he assures me, upon his honour, that he found all these stems united below-ground in one root. I alleged that so extraordinary an object must have been celebrated by many of their writers. He told me that it had, and produced several examples, Philoteo, Carrera, and some others. Carrera begs to be excused from telling its dimensions, but he says, he is sure there is wood enough in that one tree to build a large palace; and Massa, one of their most esteemed authors, says he has seen solid Oaks upwards of forty feet round; but adds, that the size of the Chestnut-trees was beyond belief, the hollow of one of which, he says, contained three hundred sheep, and thirty people had often been in it on horseback. I shall not pretend to say, that this is the same tree he means, or whether it was ever one tree or not. There are many others that are well deserving the curiosity of travellers. One of these, about a mile and a half higher on the mountain is called, *Il Castagno del Galea;* it rises from one solid stem to a considerable height, after which it branches out, and is a much finer object than the other. I measured it about two feet from the ground, it was seventy-six feet round. There is a third, called *Il Castagno del Nave*, that is pretty nearly of the same size. All these grow on a rich soil, formed originally, I believe, of ashes thrown out of the mountain."

When Houel visited it, it was in a state of decay. A house was erected in the interior, with an oven, in which, according to the custom of the country, they dried chestnuts, filberts, and other fruits which they wished to preserve for winter use ; using as fuel, when they could find no other, pieces cut with a hatchet from the interior of the tree.

M. Brunner is of opinion that the five stems of which the tree is composed, have always been distinct trunks, proceeding from one root, and that they grew in this manner in consequence of the original main trunk having been cut down, according to a custom prevalent in Sicily, before it had attained a great size, after which its place was supplied by young shoots thrown out just above the root, so that in reality the five stems are as many trees proceeding from a common origin.

THE HORSE CHESTNUT.

THE HORSE CHESTNUT.

ÆSCULUS HIPPOCASTANUM.

Natural Order—HIPPOCASTANEÆ.

Class—HEPTANDRIA. *Order*—MONOGYNIA.

The features presented by this tree are so decidedly different from those of the ordinary tenants
of our woods and forests, that a mere glance is sufficient to assure us that where the Oak, the Ash,
and the Elm are types of native trees, the Horse
Chestnut must be an alien, gladly admitted though
it be to a participation in all the privileges which
we accord to our undoubted and most highly-prized
native trees. Whether adding with its massive foliage to the deep shadow of a wood, decorating the
slopes of a park, or uniting its broad leaves to form
the canopy of a stately avenue, it is everywhere in
place, and everywhere worthy of admiration.

In early spring, before

> "The palms put forth their gems, and every tree
> Now swaggers in her leafy gallantry,"

the Horse Chestnut has given full and timely notice of the change which is in preparation. Every
one of its stout twigs is terminated by a turgid
bud hastening to anticipate its fellows in throwing
off the wintry covering, like a lusty infant strug-
gling to be released from the arms of its nurse.
After the lapse of a week, the ground is strewed
with the party-coloured scales which well did their

II.   D

duty in protecting the young shoot from the frosts
and nipping winds of February; and though the
air be motionless, others are still dropping all
around, proving that these integuments are not
passively scattered by the wind, but cast off by the
living, active, energy of the awakened bud. This
is the time beyond all others to wander in the
woods, and to be reminded by the promise afforded by the bursting bud, that we live not by the
bread which we eat, nor by the raiment that we
wear, but by the fostering Providence of God. It
is He that commanded the tender leaf to lie curiously folded in its gummy cell, till He should see
fit to call it forth—it is He that is strewing around
us the out-worn clothing of the now vigorous
shoot—the delicate blade of wheat is, in like manner, under His providential care, progressing at
the same time, more slowly, but not less surely,
towards the full ear, which is to be strengthened
by His rain, and ripened by His sun; and in every
one of the millions of puny grains which He will
enable us to gather into our barns, He is, with unerring wisdom, storing up whatever may be most
conducive to our health and temporal welfare, with
the further end more especially in view of reminding us, that "man doth not live by bread alone,
but by every word that proceedeth out of the
mouth of God."

Although we may, if we choose, learn this
lesson, at all seasons from every leaf and every
fragment of a leaf that comes in our way, yet its
truth never more decidedly thrusts itself on our
notice than when an opportunity is presented to
us of watching the unfolding buds of the Horse
Chestnut. On this account, even if it possessed no

other claims on our attention, which is far from being the case, it well deserves to be considered as a favourite tree, and such it is with most people.

The Horse Chestnut is a native of Asia, probably of northern India, whence it was introduced into Europe about the middle of the sixteenth century. In the year 1588, there was a growing specimen at Vienna, which had been planted there twelve years before, but which had not then flowered. This is said to have been imported from England, whither it had been brought from the mountains of Thibet in 1550. In France it was first raised from seed procured from the Levant, in the year 1615. Gerard, in 1579 speaks of it as a rare foreign tree; and how little it was known even by those who had seen it, may be inferred from the fact that Parkinson in 1629 places it as a fruit-tree between the Walnut and the Mulberry, and says also that it is of as good use as those trees for the fruit, which is of a sweet taste, roasted and eaten, as the ordinary sort. Some of the trees planted at Baden in the sixteenth century are said to be still in existence. There was until very recently, an avenue of splendid trees, planted in the seventeenth century, near the Royal Military Asylum at Chelsea; within the last two years, however, they have, with questionable taste, been cut down to make room for a promenade and a younger plantation.

The name Æsculus, from *esca*, food, was applied originally to a species of Oak which, according to Pliny, was highly prized for its acorns, but how it came to be transferred to the Horse Chestnut is very uncertain; perhaps, as Loudon suggests, it was given ironically because its nuts bear

a great resemblance externally to those of the
Spanish Chestnut, but are unfit for food. *Hippocastanum* is a translation of its modern name,
which was given "from its curing horses broken-
winded and other cattle of coughs." *

The Horse Chestnut is a tree of large size, frequently reaching a height of fifty or sixty feet,
with an erect trunk and a broad pyramidal outline. It may be readily distinguished even in
the depth of winter by its unusually large buds,
set on the extremities of thick and heavy-looking
branches, which are evidently destined to bear a
weighty tuft of foliage and leaves. A celebrated
German naturalist detached from this tree, in the
winter season, a flower-bud, not larger than a pea,
in which he could reckon more than sixty flowers.
The external covering was composed of seventeen
scales, cemented together by a gummy substance,
and protecting from moisture the down which
formed the internal covering of the bud. Having
carefully removed both the scales and down, he
discovered four leaves surrounding a spike of
flowers, and the latter so clearly visible, that with
the aid of a microscope, he not only counted sixty-
eight flowers, but could discern the pollen of the
stamens, and perceive that some was opaque, and
some transparent. It would be more advisable
for the young student to gather one of these buds
in the early spring, when the sun is just beginning
to melt away the gum with which the scales are
sealed together.

As the sun begins to gain power the gummy
covering of the bud melts and yields to the expanding pressure from within, when the scales

---

* Evelyn.

drop off, and the delicate green leaves are rapidly
unfolded, encircling a conical mass of embryo
flowers. In this stage the leaves present a singular
appearance, drooping with their points towards the
ground, as if not strong enough to assume a horizontal position. The buds, it has been already stated,
expand very early in spring, but by no means
prematurely, for within three or four weeks of
their first unfolding they have attained their full

LEAF AND FLOWER-BUDS OF HORSE CHESTNUT.

length, amounting sometimes to eighteen inches.
The leaves and flower-buds continue to increase
in size until May, when the latter expand; and

now, the tree having reached the meridian of its glory, stands forth prominently in all the gorgeousness of leaf and blossom. The downy covering, which was observable on the leaves in their early

FLOWER AND SEED OF HORSE CHESTNUT.

stage, has now disappeared, and they have assumed instead a rich, full green. Each leaf is composed of seven broad leaflets, unequal in size, which radiate from a common centre, a character of foliage different from that of any other British tree. Its clusters of irregular blossoms, snowy-white, dashed with pink and yellow, and affording thus early in the season a rich banquet to the venturesome bee, proclaim that the flower-bearing season now reigns paramount. "In the following description of it in this stage," says Strutt, "we can scarcely wish for anything to be altered. 'On reaching, we cannot choose but pause before, this stately Chestnut-tree, the smooth stem of which rises from the earth like a dark-coloured marble column, seemingly placed there by art to support the pyramidal fabric of beauty that surmounts it. It has just put forth its first series of rich fan-like leaves, each family of which is crowned by its splendid spiral flower; the whole, at this period of the year, forming the grandest vegetable object that our kingdom presents, and vieing in rich beauty with any that Eastern woods can boast. And if we could reach one of those flowers to pluck it, we should find that the most delicate of the fair ones of the garden or green-house do not surpass it in elaborate pencilling and richly-varied tints. It can be likened to nothing but its own portrait painted on velvet.'"

This being the only common tree in Britain of large size which bears conspicuous flowers, it has received several popular names derived from that fact, such as gigantic hyacinth, lupine-tree, giant's nosegay: this last name in particular suggests a correct notion of its vastness and showy appearance. The flowers, though exquisitely beautiful so long as they continue in perfection, soon become tarnished, and the tree consequently loses much of its grace, yet it is still a fine tree, readily distinguished at a considerable distance by its tiers of large and massive foliage. Out of the numerous flowers contained in every bunch, a few only mature their fruit, the rest drop off soon after they have begun to lose their beauty. The seed-vessels, which are prickly, or rather thorny, attain their full size in October, when they fall off, and, splitting with even-edged valves, disclose three cells, in each of which is contained a round-ish polished nut, resembling the sweet Chestnut in colour, but not, like it, terminating in a point. It rarely happens that all three nuts are perfected, frequently only two are developed, but the rudi-ments of all may be discovered.

The Horse Chestnut is one of the first trees to remind us of the approach of winter, for its leaves begin to change colour in July, and very soon to fall, but, as if to atone for this defect, its buds des-tined to expand in the succeeding spring have made so great advance as to be already conspi-cuous objects. The poet evidently had this tree in his eye, when he said that the providence of God,

> " In its case
> Russet and rude, folds up the tender germ,
> Uninjured, with inimitable art ;
> And ere one flowery season fades and dies,
> Designs the blooming wonders of the next."

As an ornament to the landscape, we have seen that the Horse Chestnut, when attired in its spring drapery, is unrivalled; it yet remains that we should consider its claims to the possession of picturesque beauty. In this respect, Gilpin, the great authority on such subjects, pronounces an opinion far from laudatory, as indeed we might naturally expect; for the particular beauties of

the Horse Chestnut are not such that they could
be represented with effect, if at all, in a picture.
It is then as an ingredient in a picture, not as a
feature of a real landscape, that he declares it to
be "a heavy, disagreeable tree. It forms its
foliage generally in a round mass, with little appearance of those breaks which we have so often
admired, and which contribute to give an airiness
and lightness, at least a richness and variety, to
the whole mass of foliage. The tree is, however,
chiefly admired for its flower, which in itself is
beautiful; but the whole tree together in flower is
a glaring object, totally unharmonious and unpic-
turesque. In some situations, indeed, and among
a profusion of other wood, a single Chestnut or
two in bloom may be beautiful. As it forms an
admirable shade, it may be of use, too, in thicken-
ing distant scenery, or in screening an object at
hand; for there is no species of foliage, however
heavy, nor any species of bloom, however glaring,
which may not be brought, by some proper contrast, to produce a good effect." Whatever truth
there may be in this criticism, it is certainly overstated, and the remarks which we have made on
the Beech,* another of the trees which the same
author is pleased unequivocally to condemn, will
equally apply in this case. One picturesque
beauty it certainly does possess, and that in a
high degree: its massive and luxuriant foliage
casts a rich, deep shadow which is eminently
beautiful in bright, sunny weather. Hence an
avenue of full-grown Chestnuts, if the ground be
not too even, and the trees planted not in an
exact line, is not without attraction even to an

* Vol. i. p. 323.

artist. Such an avenue stands in the park at
Mount Edgecumbe in Devonshire, and is by no
means the least interesting among the many beautiful landscapes which there meet the eye at every
turn.

The nuts of this tree, though not deleterious,
are unfit for human food, being very bitter. They
may, however, be applied to so many useful purposes, that it is strange they are so much neglected. Their medicinal efficacy on the animals
from which the tree takes its name, requires
confirmation; yet they are excellent food for
deer, so that, where these animals are kept,
Horse Chestnuts might be planted in numbers
with great advantage. A writer in the Gardener's
Chronicle for 1843, states, that they form a very
nourishing food for sheep. "Whilst," he says,
" I was at Geneva, in the autumn of 1837, I observed every one collecting carefully the fruit of
the Horse Chestnut, and on enquiry, I learnt that
the butchers and holders of grazing stock, bought
it readily at a certain price per bushel. I en-
quired of my butcher, who himself kept a very
extensive grazing farm, and he told me it was
given to those sheep in particular that were fat-
tening. The Horse Chestnuts were well crushed,
something in the way, so I understood, that apples
are, previously to cider being made. In Switzerland they are crushed or cut up in a machine,
kept solely for that purpose; then about two
pounds' weight is given morning and evening to
each sheep, who eat the food greedily; it must be
portioned out to them, as too much would dis-
agree with them, it being of a very heating nature.
The butcher told me it gave an excellent rich

flavour to the meat. The Geneva mutton is noted for being as highly flavoured as any in England or Wales."

They are sometimes boiled and given to poultry. Like the fruit of many other trees belonging to the same Natural Order, they contain a saponaceous principle, and when decayed they turn to a jelly, which has been found to answer the purpose of soap. Reduced to a powder and mixed with a third of flour, they are found to make better paste than that composed of flour alone. In Ireland they are used to whiten flaxen cloth, and for this purpose are rasped into water, in which they are allowed to macerate for some time. During the scarcity of 1847, it was suggested that a great saving of flour might be effected by using the starch which may be prepared from these nuts, as a substitute for wheaten starch in the process of glazing calico; but I am not aware whether or not the suggestion was acted on. M. Vergnaud has published a pamphlet, in which he proposes to convert the extracted starch into sugar, and employ it in distillation.

The Horse Chestnut will grow in most situations, but prefers a rich loamy soil. Here it grows with great rapidity; Martyn mentions some raised from the nut, that, at twelve or fourteen years of age, were covered with flowers, and were big enough to shade several chairs with their branches. A peculiarity of their growth, noticed by Hunter, is, that as soon as the leading shoot is come out of the bud, it continues to grow so fast as to be able to form its whole summer's shoot in about three weeks or a month's time. After this, it grows little more in length, but thickens and

becomes strong and woody, and forms the buds
for the next year's shoots. Owing to this rapid
rate of growth, its timber is soft, and unfit for any
use where strength and durability are required.
It is said, however, to be suitable for water-pipes
which are to be kept constantly under ground.
The bark, which is very bitter, is employed for
tanning, and also for dyeing yellow, and it has been
used medicinally as a substitute for Jesuit's bark.

The finest living specimen of Horse Chestnut in
England, is said to be that at Nocton, in Lincoln-
shire; this is fifty-nine feet high, and extends over
a space 305 feet in circumference; the branches
are so large as to require props; hence, seen from
a little distance, it resembles a Banyan-tree. At
Dawick, the seat of Sir John Nasmyth, near
Peebles in Tweeddale, are two remarkably fine
trees, which stand about twelve feet apart from
one another, and unite their foliage so as to form
a single head. Sir T. D. Lauder states, that these
are the largest in Scotland, if not in Britain.
They measure severally, sixteen and a half, and
twelve and a half feet in girth, and are in all
probability nearly 200 years old.

The following fanciful, but graphic description
of the Horse Chestnut, inserted in the Magazine
of Natural History, is from the pen of Mr. Do-
vaston.

" It was now the middle of May; the trees had
fully put forth their bright fresh leaves, and the
green boughs were luxuriant in a profusion of
flowers. We had travelled through a fine country,
when, descending the slope of a wooded valley,
we were struck with delight and admiration at a
tree of extraordinary appearance. There were

several of the sort, dispersed, singly and in groups, over the plains and grassy knolls. One, we shall attempt to describe, though well aware how feeble is the most florid description to depict an idea of so remarkable an object. In height, it exceeded 50 feet, the diameter of its shade was nearly 90 feet, and the circumference of the bole 15 feet: it was in full leaf and flower, and in appearance at once united the features of strength, majesty, and beauty; having the stateliness of the Oak, in its trunk and arms; the density of the Sycamore, in its dark, deep, massy foliage; and the graceful featheriness of the Ash, in its waving branches, that dangled in rich tresses almost to the ground. Its general character as a tree was rich and varied; nor were its parts less attractive by their extreme beauty when separately considered. Each leaf was about eighteen inches in length; but nature, always attentive to elegance, to obviate heaviness had, at the end of a very strong leaf-stalk, divided it into five, and sometimes seven leafits, of unequal length, and very long oval-shape, finely serrated. These leafits were disposed in a circular form, radiating from the centre, like the leaves of the Fan-Palm, though placed in a contrary plane to those of that magnificent ornament of the tropical forests. The central, or lower leafits, were the largest, each of them being ten inches in length, and four inches in breadth, and the whole exterior of the foliage being disposed in an imbricated form, having a beautifully light and palmated appearance. The flowers, in which the tree was profuse, demanded our deep admiration and attention; each group of them rose perpendicularly from the end of the young shoot, and was in length fourteen inches, like a gigantic hyacinth, and quite as beautiful, spiked to a point, exhibiting a cone or pyramid of flowers widely separate on all sides, and all expanded together, principally white, finely tinted with various colours, as red, pink, yellow and buff, the stamens forming a most elegant fringe amidst the modest tints of the large and copious petals. These feathery blossoms, lovely in colours, and stately in shape, stood upright on every branch all over the tree, like flowery minarets on innumerable verdant turrets.

" The natives informed us that the fruit ripens early in autumn, and consists of bunches of apples, thinly beset with sharp thorns, each, when broken, producing one or two large kernels, about two inches in circumference, of the finest bright mahogany colour without, and white within; that the tree is deciduous, and that the foliage just before its fall, changes to the finest tints of red, yellow, orange, and brown. When divested of its luxuriant foliage, the buds of the next year appear like little spears, which, through the winter, are covered with a fine glutinous gum, evidently designed to protect the embryo shoots within from the severe frosts of the climate, and which glisten in the cold sunshine like diamonds. It has the strange property of performing the whole of its vigorous shoot, more than a foot long, in the short space of three weeks, employing all the rest of the year in converting it into wood, adding to its strength, and varying its beauty. The wood when sawn is of the finest snowy whiteness. The tree is easily raised, indifferent as to soil, climate or situation; removed with safety, of quick growth, thrives to a vast age and size;

subject to no blight or disease; in the earliest
spring, bursting its immense buds into that
vigour, exuberance, and beauty, which we have
here feebly attempted to describe. The natives
said it was originally brought from the east of
Asia but grows freely in any climate, and in their
tongue its name is designated by a combination
of three words, signifying, separately, a noble
animal, an elegant game, and a delicious kernel.
Had Linnæus seen this tree, he would have as-
suredly contemplated it with delightful ecstacy,
and named it Æsculus Hippocastanum."

THE HOLLY.

THE HOLLY.

Ilex Aquifolium.

Natural Order—Ilicineæ.

Class—Tetrandria. *Order*—Tetragynia.

This " incomparable tree," as Evelyn most
justly calls it, is the most important of the English
evergreens. Whether we wander in the woods,
when all is bare and stark save the trunks of trees,
which are clothed with the borrowed verdure of
the Ivy, and save the dark but cheerful array of
armed leaves presented by the Holly; or whether,
in the bright leafy days of summer, we detect it
far off in the depths of the forest, reflecting light
from its polished mail, as brilliantly as if every
leaf were a mirror,—at any season we should be
sorry to miss it from our woodlands. But, wel-
come as the Holly is at all seasons, it belongs
more particularly to winter, for then the bright,
joyous appearance of its crimson berries, which
from our earliest years have been associated in our
minds with the festivities of Christmas, render
the tree doubly conspicuous. In one respect,
what may be said of the Hawthorn is true also
of the Holly; both these trees are emblematical
of the season in which they are most beautiful,
for it is quite as common to hear the Holly called
" Christmas," as the Hawthorn " May." Indeed
its ordinary name appears to point to the use
to which, from a very early period, it was applied,

II.   E

namely, the decoration of sacred places at the
Holy season of Christmas; for Dr. Turner, our
earliest writer on plants, calls it " Holy " and
" Holy-tree," and the same mode of spelling is
observed in a MS. ballad of yet older date, in
the British Museum.*

The origin of this beautiful custom is uncertain.
Some have supposed it to be derived from a custom observed by the Romans, of sending boughs,
accompanied by other gifts, to their friends during
the festival of the Saturnalia. This method of
shewing goodwill being at least harmless, it has
been conjectured that the early Christians adopted
it in order to conciliate their Pagan neighbours.
In confirmation of this opinion, Bourne cites a
subsequent edict of the Church of Bracara,† forbidding Christians to decorate their houses at
Christmas with green boughs at the same time
with the Pagans; the Saturnalia commencing
about a week before Christmas. Dr. Chandler
supposes the custom to have been derived from the
Druids, who, he says, decorated dwelling-places
with evergreens during winter, " that the silvan
spirits might repair to them, and remain unnipped
with frost and cold winds, until a milder season
had renewed the foliage of their darling abodes."
Certainly the custom, whencesoever it was derived,
was sanctioned by the Church; for in old Church
calendars Christmas-eve is marked, " Templa ex-
ornantur," " Churches are decked." Now, when
we recollect that of the three great Jewish festivals,

* " Holy hath berys as red as any rose."
† Non liceat iniquas observantias agere Kalendarum et ociis vacare
Gentilibus, neque lauro, neque viriditate arborum cingere domos.
Omnis enim hæc observatio Paganismi est."—*Brac. Can.* lxxiii.

namely, the Passover, Pentecost, and the Feast
of Tabernacles, the two former were undoubtedly
typical of the Christian festivals Easter and
Whitsuntide, and were, if I may be allowed the
expression, merged in them, may we not infer
that the early Christians adopted the custom of
decking their churches and dwellings with green
boughs to shew the connexion between the Jewish
Feast of Tabernacles, and the festival at which
they commemorated the fact that, "the Word
was made flesh and dwelt," or, as it may be more
correctly rendered, "*tabernacled* among us?" In
the absence of all evidence, this conjecture appears
to be quite as consistent with reason as any of
the others which have been made, and certainly
more in accordance with the piety of the early
Christians. In some rural districts, the thorny
leaves of the Holly, and its scarlet berries, like
drops of blood, are thought to be symbolical of our
Saviour's sufferings; for the same reason, perhaps,
in the language of several of the northern countries
of Europe, the tree is called "Christ's thorn."

Mr. Knapp, in his instructive and entertaining
"Journal of a Naturalist," makes the following
remarks on the abuses of this custom : "Christ-
masing, as we call it, the decorating of our
churches, houses, and market meats with evergreens, is yet retained among us; and we, growers
of such things, annually contribute more than we
wish for the demand of the towns. Sprays and
sprigs may be connived at, but this year I lost most
of my beautiful young Holly-trees, the cherished
nurslings of my hedgerows. The Holly, though indigenous with us, is a very slow growing tree,
and certainly the most ornamental of our native

foresters. Its fine foliage shining in vigour and health, mingling with its bright coral beads, gives us the cheering aspect of a summer's verdure, when all besides is desolation and decay. It

HOLLY BERRIES—WINTER OF 1845-6.

is not only grateful to the eye, but gives us pleasure when we contemplate the food it will afford our poor hedge-faring birds, when all but its berries and those of the Ivy are consumed; and we are careful to preserve these gay youths of promise, when we trim our fences; but no sooner do they become young trees in splendid beauty, than the merciless hatchet on some December's night lops off all their leaves, leaving a naked unsightly stake to point out our loss: and we grieve and are vexed, for they never acquire again comparative beauty."

On the custom of decorating dwelling-houses with Holly at Christmas appears to have been grafted another, that of replacing them at other seasons, with a succession of different boughs.

CEREMONIES FOR CANDLEMAS-DAY.

" Down with Rosemary and Bays,
Down with the Mistletoe ;
Instead of Holly now upraise
The greener Box for show.

" The Holly hitherto did sway ;
Let Box now domineer,
Until the dancing Easter-day
Or Easter's eve appear.

" Then youthful Box, which now hath grace
Your houses to renew,
Grown old, surrender must his place
Unto the crisped Yew.

" When Yew is out, then Birch comes in,
And many flowers beside,
Both of a fresh and fragrant kin
To honour Whitsuntide.

" Green rushes then, and sweetest bents,
With cooler oaken boughs,
Come in for comely ornaments,
To re-adorn the house.

Thus times do shift ; each thing his turn doth hold ;
New things succeed as former things grow old."
HERRICK'S *Hesperides.*

This tree was formerly known by the names of Hulver and Holme, besides its more usual appellation. It is still called Hulver in Norfolk, and Holme in Devonshire, in which last county it has

given the name of Holme Chase to a beautiful part
of Dartmoor, where it abounds. Evelyn says that
the vale near his house, Wotton in Surrey, was
anciently called Holmsdale, for the same reason.

Pliny describes the Holly under the names of
Aquifolium and Aquifolia, that is, needle-leaf,
and adds, that it was the same with the tree
called, by Theophrastus, Cratægus, a statement
which the commentators pronounce erroneous,
the Greek name for the Holly being Agria. He
says also, that if planted in a house or farm, it
repels poison, and that its flowers cause water
to freeze. A staff of its wood, he adds, if thrown
at any animal, even if it fall short of the mark,
has the wonderful property of compelling such
animal to return and lie down by it.

Phillips, in his *Sylva florifera*, professes to quote
from Pliny a description of certain ancient Hollies
which stood near the city of Tibur, in Italy: but
his version of the passage is erroneous, for the
tree mentioned is the Ilex or Evergreen Oak, and
not the Holly. Loudon, following Phillips, falls
into the same mistake; and, strangely enough, in
another part of his Arboretum describes the same
trees again, bearing their right name.

The Holly is a native of most of the central
and southern parts of Europe, but is said nowhere to attain so great a size as in Great Britain,
where it sometimes ranks as a second-rate forest
tree. As it grows very slowly, if it were im-
patient of the drip of other trees we should
never see it in our woods; but the Divine Power
which fixed its rate of growth, ordained at the
same time that it should thrive best under the
shade of its lofty companions. Hence we fre-

quently see it deepening the gloom of a forest, where it is rarely visited by even a few straggling sunbeams, and where the only moisture which bathes its leaves is derived from the superfluous rain which has dripped from the overshadowing foliage of its more elevated comrades. When planted among trees which are not more rapid in growth than itself, it is sometimes drawn up to a height of fifty feet or more. In Wistman's Wood, Dartmoor, (described in volume i. page 14,) there stands among the stunted Oaks, which have been unable to make head against the sweeping moorland blast, a single Holly which overtops by many feet every tree in the wood; but these are rare instances. More frequently it is contented with the humble elevation of thirty feet, or even less, sometimes forming a perfect pyramid, leafy to the base, at other times sending up a clean stem furnished with a bushy head. The bark is of a remarkably light hue, and is very liable to be invested with an exceedingly thin lichen,* the fructification of which consists of numerous curved black lines closely resembling oriental writing. The leaves are thick, tough, and glossy, and edged with stout prickles, of which the terminal one only is invariably in the same plane with the leaf. In May, the Holly

OPEGRAPHA
SCRIPTA.

* *Opegrapha scripta.*

bears in the axils
of the leaves crowd-
ed, small, whitish
flowers, which are
succeeded by the
brilliant coral ber-
ries too familiarly
known to require
description.   The

HOLLY IN BUD—WINTER OF 1846-7.

same tree rarely produces abundant crops of
flowers in consecutive seasons; consequently, if
we find a Holly one winter loaded with berries,
in all probability it will bear but few in the

FLOWERS OF THE HOLLY.

following winter, but we shall discover in the
clusters of unexpanded buds, ample intimation
of the abundant crop which it intends to produce in the succeeding year. Another pecu-

liarity of the Holly was observed, and thus beautifully moralized on by the poet Southey :

"O Reader! hast thou ever stood to see
            The Holly-tree?
The eye, that contemplates it well, perceives
            Its glossy leaves
Ordered by an Intelligence so wise,
As might confound the atheist's sophistries.

" Below a circling fence its leaves are seen,
            Wrinkled and keen ;
No grazing cattle through their prickly round
            Can reach to wound ;
But as they grow where nothing is to fear,
Smooth and unarmed the pointless leaves appear.

" I love to see these things with curious eyes,
            And moralize:
And in this wisdom of the Holly-tree
            Can emblems see,
Wherewith, perchance, to make a pleasant rhyme,
One which may profit in the after-time.

" Thus, though abroad perchance I might appear
            Harsh and austere,
To those who on my leisure would intrude
            Reserv'd and rude ;
Gentle at home amid my friends I 'd be,
Like the high leaves upon the Holly-tree.

" And should my youth, as youth is apt, I know,
            Some harshness show,
All vain asperities I day by day
            Would wear away,
Till the smooth temper of my age should be
Like the high leaves upon the Holly-tree.

" And as when all the summer leaves are seen
            So bright and green,
The Holly-leaves their fadeless hues display,
            Less bright than they—
But when the bare and wintry woods we see,
What then so cheerful as the Holly-tree ?—

"So serious should my youth appear among
    The thoughtless throng;
So would I seem amid the young and gay
    More grave than they;
That in my age as cheerful I might be
As the green winter of the Holly-tree."

The leaves of the Holly remain attached to the tree for several years, and, when they have fallen,

for a long time defy the action of air and moisture. The following lines must have been written by a close observer of nature :—

"Where leafless Oaks towered high above
I sate within an under grove
Of tallest Hollies, tall and green ;
A fairer bower ne'er was seen.
From year to year, the spacious floor
With withered leaves is covered o'er,
You could not lay a hair between :
And all the year the bower is green.
But see, where'er the hailstones drop,
The withered leaves all skip and hop,

> There 's not a breeze—no breath of air—
> Yet here—and there—and everywhere—
> Along the floor, beneath the shade,
> By these embowering Hollies made,
> The leaves in myriads jump and spring,
> As if with pipes and music rare,
> Some Robin Goodfellow were there,
> And all those leaves, in festive glee
> Were dancing to the minstrelsy."
>
> WORDSWORTH.

The leaf, having a very tough and durable fibre, takes a long while to decay, and may frequently be picked up, a frame filled in with network entirely divested of cuticle.

As an ornament of the landscape, Gilpin considers the Holly to be a tree of singular beauty, especially in forests, where, he says, " mixed with Oak or Ash, or other trees of the wood, it contributes to form the most beautiful scenes; blending itself with the trunks and skeletons of the winter, or with the varied greens of summer. In many situations it appears to great advantage, but particularly growing round the stem, as it often does, of some noble Oak on the foreground, and filling up all the space to his lower boughs. In summer it is a fine appendage; and in autumn its brilliant leaf and scarlet berry make a pleasing mixture with the wrinkled bark and hoary moss and auburn leaves of the venerable tree which it encircles." The Holly is scarcely less ornamental in the shrubbery and lawn than in the wild woodland, for in summer its foliage contrasts well with that of every other tree and with the turf on which it rises; and in winter, when the greater part of the vegetable world is content for a while to forego its garniture, it stands out pre-eminently conspicuous and beautiful. When the ground is covered with snow, it is a brilliant object, and is made all the more interesting by the multitudes of birds which flock to it for the sake of its berries. Evelyn is lavish in his praises of the Holly, though his eulogium is strongly tinctured with the taste of the age in which he wrote:—" Is there under heaven a more glorious and refreshing object of the kind, than an impregnable hedge of about four hundred feet in length, nine feet high, and five feet in diameter, which I can shew in my now ruined gardens at Say's Court (thanks to the Czar of Muscovy,*) at any time of the year, glittering with its armed and varnished leaves ? The taller standards at orderly distances, blushing with their natural coral, it mocks the rudest assaults of the weather, beasts, or hedgebreakers." This description will perhaps be deemed over-coloured ; nevertheless, the Holly is far superior to any other tree we possess for the formation of living hedges. It is very durable, patient of any amount of clipping, liable to no blight, and equally impenetrable at all seasons of the year. It is, besides, proof against the most inclement weather, and may be trained to a greater height than any other hedge-tree. At Tyningham, the seat of the Earl of Haddington, there are 2952 yards of Holly-hedge, chiefly planted in 1712. In height they vary from ten to twenty-five feet, and are from nine to thirteen feet wide at the base. They are regularly clipped every April,

---

* The Czar, Peter the Great, resided at Mr. Evelyn's house during his stay in England, in order to be near the Dockyard at Deptford. The Emperor, it is said, took very great delight in the amusement of being wheeled in a barrow through the thick Holly-hedges, which were the pride of the garden.

and are protected from cattle and other injury by a ditch on either side. In Evelyn's time the art of the nurseryman was scarcely known, and he recommends that they should be picked out of the woods while young, before they have been cropped by sheep, and when planted where they are destined to remain, to be "shorn and fashioned into columns and pilasters, architectionally shaped, and at due distance; than which nothing can be more pleasant, the berry adorning the intercolumniations with scarlet festoons." Now, however, that they may be purchased at seven shillings a thousand, few will think it worth while to adopt the laborious process which he recommends: "architectional" columns have long ago disappeared, and given place to a style of gardening more in accordance with the proprieties of nature.

The Holly will grow in almost any soil, provided it is not too wet, but attains the largest size in rich, sandy loam. The most favourable situation seems to be a thin scattered wood of Oaks, in the intervals of which it grows up, at once sheltered and partially shaded.

Beckmann, in his history of the discovery of alum, relates the following incident:—John di Castro, son of the celebrated lawyer Paul di Castro, had an opportunity at Constantinople, where he traded in Italian cloths and sold dyed stuffs, of making himself acquainted with the method of boiling alum. He was there at the time when the city fell into the hands of the Turks (1453); and after this unfortunate event, by which he lost all his property, he returned to his own country. Pursuing there his researches

in natural history, he found in the neighbourhood of Tolfa great abundance of Holly, a tree
which he had also observed growing plentifully
in the aluminous districts of Asia; from this he
conjectured that the earth of his native soil might
also contain the same salt; and he was confirmed
in that opinion by its astringent taste. His
discovery being made known to Pope Pius II.
led to the establishment of the first alum-works
in modern Europe. The first alum-works in
England were erected at Gisborough, in York-
shire, in the reign of Queen Elizabeth, the existence of the mineral having been discovered by
Sir Thomas Chaloner, who, observing that the
leaves of trees were here of a lighter green than
in other places, suspected that the difference was
owing to the presence of some peculiar mineral
substance, and found on examination that the soil
abounded in aluminous salt. The method of
extracting the mineral being unknown in England, he induced some of the Pope's workmen to
lend their assistance, and finally enriched himself
at the expense of no very gentle imprecations
from the sovereign monopolist. Evelyn observes,
that the Holly often indicates where coals are to
be found. The truth is, that this tree being remarkably indifferent to soil, equally indicates the
proximity of alum, coal, and most other minerals.

The trunk of the Holly, like that of the
Beech, frequently has small round knots attached
to it; these are composed of a smooth nodule of
solid wood embedded in bark; they may readily be
separated from the tree by a smart blow. Hollies
are not unusual which, although they grow most
luxuriantly and produce abundance of flowers,

never mature berries. This barrenness is occasioned by an imperfection in the pistil, the
cause of which is unknown.

The cultivated varieties of Holly are very
numerous; of these three are distinguished by
the unusual colour of their berries, yellow, black,
and white ; the rest are characterized by their variegated foliage, or by the presence of a larger
number of prickles than ordinary. Of later years
more attention has been paid to the discovery of
new species of trees than to the cultivation of new
varieties, in consequence of which many of the
sorts mentioned by old authors are now extinct.
Of all variegated trees, the Holly is the most
pleasing, the yellow tint of its foliage being
generally of a bright decided tone, and therefore
not suggestive of disease, an idea which is associated with most other trees which have their
leaves blotched with yellow. In winter, when
flowers are scarce, the garden and shrubbery are
much indebted to the more showy varieties for
the double contrast afforded by their leaves and
berries. They are propagated by grafting on the
common sort, and attain an equal size. Hayes
mentions one at Ballygannon in Ireland twenty-
eight feet high, with a trunk five feet in circumference ; at Enys in Cornwall is another, the
stem of which is three feet and a half in circumference, and fifty-nine feet high. This last is
associated with other slow-growing trees; hence
it has been drawn up to this unusual height.

The uses of the Holly in its natural state are
scarcely worth notice. Deer will eat the leaves in
winter, and sheep thrive on them. The berries
are violently emetic; but a decoction of the

bark is said to allay a cough. Rats and mice
occasionally injure the young trees by knawing the
bark of them, especially when the ground is
covered with snow. It is infested by but few
insects: the azure-blue butterfly (*Polyommatus
Argiolus*) delights to hover about it, and settle on
it ; and another small insect passes the larva and
pupa stages of its existence between the upper
and under cuticle of the leaf; but, with these
exceptions, it is exempt from insect depredations.

Paterson, in his "Insects mentioned in Shakspeare," speaks of the Holly as being occasionally infested to a great degree with honey-dew. He once
noticed one at Castle Willan, in Ireland, on which
a number of wasps were continually alighting, running rapidly over its leaves, and flitting from
branch to branch. A great many Hollies were
scattered over the lawn, but not one exhibited the
same lively bustle. On a close examination, he
found that the wasps were not its only visitors.
A number of ants were plodding quietly along
the twigs and leaves, exhibiting by their staid and
regular deportment, a singular contrast to the
rapid and vacillating movements of the wasps.
The object of attraction to both was the honey-
dew deposited by the aphides which infested the
tree. Ants, however, do not in all cases frequent
trees for the sake of supplying themselves with
food from this source. In Guiana, grows a tree
(*Triplaris Americana*) called by the natives Jacina,
or Ant-tree. It attains the height of fifty or
sixty feet, and at certain seasons presents a most
attractive appearance. The inconspicuous flowers
fall off soon after their expansion, and the calyx

then enlarges and becomes white, with a tint of vermilion. But woe to the incautious botanist who attempts to pluck them. The trunk and branches are hollow, with openings at intervals, and are inhabited by numerous light brownish ants, about a third of an inch in length, which inflict the most painful bites. They fall on their prey with the greatest virulence, burying their sharp mandibles in any soft substance which presents itself, and emitting a whitish fluid: their bite causes swelling and itching for several days. If placed in confinement, they attack and kill one another.

The wood of the Holly is hard, compact, and of a remarkably even substance throughout. Except towards the centre of very old trees, it is beautifully white, and being susceptible of a very high polish, is much prized for ornamental ware. It is often stained blue, green, red, or black ; when of the latter colour, its principal use is as a substitute for ebony, in the handles of metal tea-pots. Mathematical instruments are also made of it, and it has even been employed in wood-engraving instead of box. The wood of the silver-striped variety is said to be whiter than that of the common kind. Of the bark, stripped from the young shoots, boiled and suffered to ferment, birdlime is made ; but the greater quantity of this substance used in England is imported from Turkey.

In the north of England it was formerly so abundant about the lakes, that birdlime was made from it in large quantities, and shipped to the East Indies for destroying insects. It is raised from seeds, which do not germinate until the second year ; hence the berries are generally buried in a heap of earth for a year previously to being sown.

One of the finest specimens of Holly now exist-ing in England, is mentioned by Grigor, as grow-ing at Spring Grove, Norfolk, it is about sixty feet high, with a bole of five and a-half feet in circumference, and twenty-five in length: but others of nearly equal dimensions are to be met with here and there.

On the glebe of St. Gluvias Vicarage, Cornwall, stands another exceedingly fine specimen. It is in form a pyramid, thirty feet in height; .the lower branches descending to the ground. It has three principal stems, measuring severally, at the surface of the ground, five feet three inches, three feet nine, and four feet four inches, equivalent to one stem seven feet nine inches in circumference. The spread of the branches is a hundred and thirty-five feet in circumference.

A low shrubby plant, which occurs not un-frequently in woods and hedges, is sometimes called Knee-holly, though in no way allied to the true Hollies. Its botanical name is *Ruscus acu-leatus*, and it is also called Butcher's Broom. It belongs to the natural order of *Liliaceæ*, and is the only indigenous shrub in the class *Endogens*.* It may easily be detected by its tough, green, striated stems, which are destitute of bark, and send out from the upper part many short branches. The rigid leaves are a mere expansion of the stem, and terminate each in a single sharp spine. The small green flowers are solitary in the centre of the leaves, and the fertile ones are succeeded by bright scarlet berries as large as cherries,

* For an explanation of this term, see Vol. i. p. xvi.

which remain attached to the plant all the winter.
The young shoots are sometimes eaten like those
of Asparagus, a plant to which it is closely allied ;

BUTCHERS' BROOM.

when matured, they are bound in bundles, and
sold to the butchers, who use them for sweeping
their blocks. The name Knee-holly appears to
have been given from its rising to about the

height of a man's knee, and from its having, like the true Holly, prickly leaves.

The most interesting species of foreign Holly is the *Ilex Paraguayensis*, the following account of which is condensed from the first volume of the London Journal of Botany.

Few persons are ignorant of the fact, that, throughout a large portion of South America, a favorite beverage is employed under the name of *Maté* or *Paraguay Tea;* but many are of opinion that the plant producing it resembles the Tea-plant of China, little aware that it is a kind of Holly, and a species not very unlike some of the varieties of our English Holly. Until about 1822, nothing whatever was known respecting the particular genus and species of the shrub whose leaves furnished the *Maté* or *Paraguay Tea.* The former of these appellations originated in the name of the cup "Maté," from which the tea was drunk. It had further the name of *Yerba,* the herb. A French Naturalist, St. Hilaire, appears to be the first European who furnished a botanical account of the plant. He found it growing abundantly in the woods of Paraguay, both in flower and fruit, and was enabled to pronounce it to be an Ilex. It forms a small upright growing tree, about fifteen feet high. The trunk is about the thickness of a man's thigh, with a shining and whitish bark, and branches, which, like those of the laurel, grow pointing upwards, and the whole plant has a tufted and much branched appearance. In those situations where the leaves are regularly gathered, it only forms a shrub, because it is periodically stripped of all its foliage and small branches every year.

The leaves are evergreen, from two to six inches
long, notched, and more or less acute, and in
some varieties dotted with minute black glands.
The principal *Yerbals*, or woods of the Yerba-
tree, are situated near the small town of Villa
Real. So impenetrable, and in many parts over-
run with brushwood, are these forests, and everywhere so tenanted with reptiles of a venomous
character, that the only animals capable of being
driven through them are oxen and mules; the
former, necessary for the food of the colony of
Yerba-makers, and the latter, indispensable to the
conveyance out of the woods of the tea, after it
has been manufactured and packed. These poor
beasts are so tortured with the bites of mosquitoes as to yell dreadfully when driven along, and
the Peons, or slaves who ride the mules, have
their legs cased in raw hides, their faces covered
with tanned sheepskin, and their hands protected
by gloves of the same material. The party generally consists of from twenty to fifty persons,
and is collected by the merchant, who has obtained permission from the governor to cut the
leaves, and who immediately notifies in public his
intention in those districts where the natives reside who best understand the business. The merchant comes, provided with goods, mules, hides,
and hatchets; and gives the persons whom he engages a certain quantity of articles in advance, on
credit. Thus equipped, they set off in the direction of the Forests of Yerba. When bivouacing
at night, a high stage is erected, fifteen feet from
the ground, whereon a roof is laid, on which the
whole colony sleep to avoid the mosquitoes, which
never rise so high in the air, and also to be safe

from the jaguars and noxious reptiles which
swarm in the forests. When they come to a
Yerbal or forest of Maté trees sufficiently large
to make it worth their while to halt and collect
the leaves, they begin by constructing a long
line of wigwams covered with the broad leaves
of the Banana and Palm, beneath the shade of
which they expect to pass nearly six months.
The next process is to prepare the piece of
ground on which the small branches, twigs, and
leaves of the Yerba are first scorched. The soil
is beaten with heavy mallets, till it becomes hard
and smooth. When this task is accomplished, the
cutters disperse singly through the woods, and
return laden with as many branches as they can
carry. The leaves, when thoroughly dried, are
carried to the prepared ground, and placed on
a kind of arch made of hurdles. A large fire
is kept up beneath, and the foliage is thoroughly
scorched without being suffered to ignite; after
which the dry platform is swept clean, and the
leaves are beaten off the branches by means
of sticks, and reduced nearly to powder. Recently, there has been substituted for this part
of the human labour a rude mill, in which the
scorched foliage and slender twigs are together
ground to powder, thus completing the process,
and rendering the Paraguay Tea fit for use. It is
then conveyed to a large shed, where it is received,
weighed, and stored, by the overseer.

The operation of packing is the most laborious
part; this is effected by cramming and beating
into a bull's hide, which is damped, and fixed
firmly into the ground, the greatest possible quantity of the pulverized Maté. From 200 to 300

pounds are often pressed into one of these leathern sacks, which is then sewed up, and left to tighten over the contents, and the heat of the sun will in two days cause the hide to shrink into a substance as hard as stone.

The mode of using the leaf thus prepared, is to infuse a portion of it in a small open vessel, called a *Maté*, and to suck it hot through a tube, called a *Bombilla*,* which is perforated at the end with small holes, to prevent the escape of the particles of leaf which are suspended in the fluid. The infusion must be made with fresh water each time, and drunk off immediately, or the liquor becomes as black as ink; but the powder will bear to be steeped at least thrice.

The whole party is supplied by passing the Maté-cup from hand to hand, the guests using their own tubes, which are made of silver, glass, or reed. Another kind of Maté-cup is furnished with a spout, which is used by the whole company indiscriminately. The superior kinds are often made of a calabash, mounted with silver, and fixed on a stand; or of silver itself, elegantly carved and chased. Some persons add to the beverage a lump of burnt sugar, or a few drops of lemon-juice.

For nearly a century and a half, an infusion of this plant has been the common and favourite drink of the settlers at Paraguay, who adopted the practice from the natives, and the custom soon extended itself to other parts of South America; so that in proportion to the population, in-

---

* The annexed woodcut contains an accurate representation of a Maté-cup, made of a hollow gourd, and a silver Bombilla, in the possession of the Rev. W. S. Hore.

no part of the world is Chinese tea more extensively used than Maté is throughout a great portion of South America,—in Brazil for example,

in Peru and Chili, and everywhere to the south of those vast territories. So devoted do the inhabitants of these countries become to its use, that to break it off, or even to diminish the customary quantity, is almost impossible. Like opium, it certainly appears to rouse the torpid, and calm the restless; but, as in the case of that noxious drug, the immoderate use is apt to occasion diseases similar to those consequent on the practice of drinking strong liquors. Persons who are fond of it consume about an ounce a day. In the mining countries, the Maté is most universally taken, and the Creoles throughout South America are passionately addicted to this beverage, and never travel without a supply of the leaf, which they infuse and imbibe before each meal, and sometimes much oftener, never tasting food unless they have first drunk their Maté. Besides the quantity consumed in Paraguay itself, nearly six millions of pounds are annually exported.

Frezier, who visited Chili and Peru in 1712, supposed Maté to be the produce of a herbaceous plant. "The herb," he says, "which produces it is called by some St. Bartholomew's herb, who, they pretend, came into those provinces, when he made it wholesome and beneficial, whereas before it was venomous; being only brought dry, and almost in powder, I cannot describe it."

The beverage itself is bitter and aromatic, and possesses a peculiar smell and flavour unlike any European herb with which I am acquainted. To most palates, accustomed to Chinese tea, it would be perhaps, at first, anything but grateful. I can readily believe, however, that the very peculiarity soon becomes agreeable, and secures for it the extensive adoption above described.

THE COMMON BIRCH.

THE BIRCH.

BETULA ALBA.

Natural Order—AMENTACEÆ.

Class—MONŒCIA.   *Order*—POLYANDRIA.

No tree is more generally or more deservedly admired on the ground of its own intrinsic beauty than the Birch. As the Oak has no tree to vie with itself in the sterner attributes of majesty, dignity, and strength, so the

> " most beautiful
> Of forest-trees, the Lady of the Woods,"

stands unrivalled in lightness, grace, and elegance. In one respect it even claims precedence over the monarch of the forest, and that one which its slender and delicate form would least lead us to expect: it stands in need of no protection from other trees in any stage of its growth, and loves the bleak mountain side and other exposed situations, from which the sturdy Oak shrinks with dismay. But the style of beauty in which each of these trees excels is so different in kind, that neither of them can properly interfere with the other.

Pliny describes the Birch as a slender tree inhabiting the cold parts of Gaul. The branches, he says, were used for making baskets and the hoops of casks ; and the fasces or bundles of rods, which were carried before the Roman magistrates

were made of birch twigs; the use, therefore, of
the weapon, which, in modern times, is the terror
of idle schoolboys, is of high classical authority.
Branches of this tree were formerly used for
decking the houses in Rogation week, as Holly
is at Christmas. Gerard says the branches " serve
well to the decking up of houses and banquetting
roomes for places of pleasure, and beautifying the
streetes in the Cross, or gang, weeke " (the same
as Rogation week) " and such like."

The Birch is a native of the colder regions of
Europe and Asia. Throughout the whole of
the Russian empire it is more common than
any other tree, being found in every wood and
grove from the Baltic Sea to the Eastern Ocean and
frequently occupying the forest to the exclusion of
almost every other tree. It grows from Mount
Etna to Iceland; in the warmer countries being
found at a high elevation in the mountains, and
varying in character according to the temperature.
In Italy, where it grows, though it appears from
Pliny's account not to have been noticed by the
ancients, it forms little woods at an elevation of
six thousand feet; on some of the Highland
mountains it is found at the height of three
thousand five hundred feet. In Greenland it is
the only tree; but wherever it grows it dimi-
nishes in size according to the decreased temperature to which it is exposed.

The peculiar characteristics of the Birch are,
as it has been remarked, lightness and elegance,
qualities which are owing to the slenderness of
the main stem in proportion to its height, the
wiriness of the branches, and the thinness and
small size of the foliage. It is equally remark-

able for the colouring of its bark, which is marked with brown, yellow, and silvery touches, which are very picturesque, contrasting well with the dark green hue of the leaves. The younger twigs have no such variety of colour, being of an uniform purple brown. The leaves are sharp-pointed, stalked, and unevenly serrated. In April and May the flowers appear in the form of droop-ing catkins, some of which produce stamens only and drop off early. The fertile ones bear very

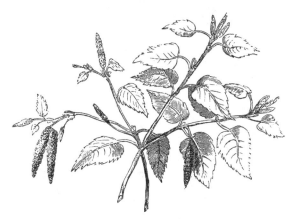

LEAF AND FLOWER OF THE BIRCH.

small winged nuts, and fall to pieces when ripe, scattering the numerous seeds. The barren cat-kins are formed in summer, but do not expand till the fertile catkins appear in the following spring. A kind of resin exudes from the leaves and young twigs, which is highly fragrant, espe-cially after rain or heavy dew. This resin ap-

pears to have been collected in Pliny's time ; as
he speaks of a *bitumen* which the tree produces.
The odour arising from it is very perceptible to a
person passing near a tree, and affords another
reason why the Birch should be planted near
houses.

THE WEEPING BIRCH.

A variety of the Birch is often met with in the
Highlands, which differs from the common species by having the shoots pendulous. This is a
yet more elegant tree than the first described, and
is frequently planted in parks and gardens. It

possesses another advantage in being of quicker
growth, and attaining a larger size.   There is also
a slight difference in the leaves, which are smaller
and somewhat downy.   Sir W. J. Hooker observes that the branches are more warty in the
Weeping Birch than in the common kind, and
Sir T. D. Lauder has noticed that the fragrant
gum or resin, which exudes from both varieties,
hardens on the spray of the Weeping Birch, and
causes a rough appearance.

   The Birch is a tree of rapid growth, especially
when young; and as it is little affected by exposure, it forms an excellent nurse for other
trees.   The soil which it prefers is turf over sand,
and in such situations it attains maturity in about
fifty years; but it seldom exceeds fifty feet in
height, with a trunk from twelve to eighteen inches
in diameter.   The bark possesses the singular
property of being more durable than the wood
which it encircles.   Of this the peasants of
Sweden and Lapland, where Birch is very abundant, take advantage, and, shaping it like tiles,
cover their houses with it.   Travellers in Lapland
have noticed in the Birch-forests, that when the
soil is very scanty, the trees are liable to be
blown down; so that, in some places, as many
are seen lying on the ground as are left standing.   Such as have lain long are found to have
entirely lost the substance of the wood, the
bark remaining a hollow cylinder without any
symptom of decay.   In North America this
durability of the bark is turned to good account.   The Canadians select a tree with a
large and smooth trunk.   In the spring two
circular incisions are made quite through the bark

several feet from each other. Two vertical incisions are then made on opposite sides of the
tree ; after which a wooden wedge is introduced,
by which the bark is easily detached. These
plates are usually ten or twelve feet long, and two
feet nine inches broad. To form the canoe, they
are stitched together with the fibrous roots of the
White Spruce. The seams are coated with resin
from the Balm of Gilead. Great use is made of
these canoes by the savages and by the French
Canadians, in their long journeys into the interior
of the country; they are very light, and are easily
carried on the shoulders from one lake or river to
another. A canoe calculated for four persons
weighs from forty to fifty pounds. Some of them
are made to carry fifteen passengers. This species
is known as the Paper Birch. The thin white
bark of the common Birch, which peels off like
paper, is highly inflammable, and will burn like a
candle.

The Birch abounds in a sweet watery sap, which
was formerly much valued for its supposed medicinal virtues. The method pursued in collecting
it is as follows :—About the beginning of March
an oblique cut is made, almost as deep as the
pith, under some wide-spreading branch, into the
which a small stone or chip is inserted, to keep
the lips of the wound open. To this orifice a
bottle is attached to collect the flowing juice,
which is clear, watery, and sweetish, but retains
something both of the taste and odour of the tree.
Various receipts are given for the preparation
of the wine. That recommended by Evelyn can
hardly fail to produce an agreeable beverage. He
directs that it should be boiled for an hour, with

a quart of honey to every gallon of juice, a few cloves, some lemon peel, and a small proportion of cinnamon and mace. It should then be fermented with yeast and bottled. This process, according to the same author, does not injure the tree, for he mentions having observed a Birch which was so treated for very many years, and nevertheless grew to an unusual size.

The wood of the Birch is white, shaded with red, and, if grown in a very cold climate, it lasts a long while. It is used for packing-cases, turnery, wooden shoes, and the felloes of wheels, but is inferior to other kinds of timber for all these purposes. A piece of birch-wood was once found in Siberia, changed entirely into stone, while the cuticle, or outer coating of the bark, of a satiny whiteness, was exactly in its natural state, perfectly well preserved. This proves what was before said of the durability of the bark. Thin pieces of the cuticle are sometimes placed between the soles of shoes, and are found to resist the wet. The bark is even wrapped round the lower end of posts, which are inserted in the ground, to prevent the moisture from penetrating them. The bark of large trees is used by the Laplanders as a kind of cloak, a hole being made in the centre to admit the head. From smaller trees, about the size of a man's leg, they make boots by removing the wood and leaving a seamless tube of bark. In seasons of scarcity, the inner bark is sometimes ground with corn, and made into bread; but this, we must hope, happens but rarely.

From the leaves a yellow dye may be prepared. The wood makes excellent charcoal for gunpowder and crayons, and in northern countries, other parts of the tree are applied to various uses, not, indeed, from any particular fitness of the Birch, but from the absence of trees of any other kind: "the Highlanders of Scotland make everything of it, they build their houses, make their beds, chairs, table, dishes, and spoons; construct their mills; make their carts, ploughs, harrows, gates, and fences, and even manufacture ropes of it. The branches are employed as fuel in the distillation of whiskey; the spray is used for smoking hams and herrings, for which last purpose it is preferred to every other kind of wood. The bark is used for tanning leather, and sometimes, when dried and twisted into a rope, instead of candles. The spray is used for thatching houses: and, dried in summer, with the leaves on, makes a good bed, where heath is scarce."—LOUDON.

In Russia an oil is extracted from the Birch, which is used in the preparation of Russian leather. For this purpose, the white bark, taken either from recent trees, or from the decayed trees which are found in the woods, is gathered into a heap, and pressed into a pit shaped like a funnel; it is then set on fire, and covered with turf.

The oil which trickles down the sides drops into a vessel placed to receive it, and is then stowed away in casks. The purest oil swims at the top, and when used for anointing leather not only imparts a fragrant odour, but makes it durable. Owing to the presence of this oil, books bound in Russian leather are not liable to become mouldy; they also prevent mouldiness in books bound in other leather which happen to be near them.

DWARF BIRCH.

The Birch is liable to a disease which shews itself by producing on the upper branches large

tufts of twigs, which, seen at a distance, resemble
crows' nests. How it originates is unknown,
some persons assigning it to the puncture of an
insect, others to a peculiarity in the soil.

A great variety of insects prey on this tree, but
not in sufficient number to do it any material
injury; and many kinds of fungi fix themselves
on the decaying wood. A species of mushroom
(*Amanita muscaria*) is found in Birch-woods
which is amongst the most poisonous of the
tribe. Taken in small quantities it is narcotic,
and produces intoxication. In Kamschatka,
where it is very plentiful in some seasons, it is
carefully collected in the hottest months, and
hung up in the air to dry. It is swallowed without being chewed, and produces no effect for an
hour or two, but at the expiration of that time
giddiness and drunkenness come on together, and
the subject of the disgusting experiment is no
longer master of his reason or his actions.

A distinct species of Birch, *betula nana*, Dwarf
Birch, is found in Scotland and in all the northern countries of continental Europe and America.
It is a low wiry shrub, rarely exceeding three feet
in height; with numerous round, notched leaves,
which are beautifully veined. By the Laplander
it is applied to the same purposes as the twigs of
the larger kind.

THE ALDER.

THE ALDER.

ALNUS GLUTINOSA.

Natural Order—AMENTACEÆ.

Class—MONŒCIA.   *Order*—TETRANDRIA.

THIS tree is botanically distinguished from the preceding by having its fertile catkins oval, and its seeds not winged, whereas the fertile catkins

FLOWER AND LEAF OF THE ALDER.

of the birch are cylindrical, and the seeds furnished with a border. Though so nearly resembling each other in the structure of the flowers as to have been placed by some botanists in the same genus, in general form, character of the foliage, and place of growth, no two trees are more distinct; for while the Birch is singularly marked by elegance of form and lightness of foliage, the Alder is stiff, heavy, and even gloomy.

The Alder is a very widely diffused tree, growing by the sides of rivers and in swampy places unfit for the growth of other trees, throughout the whole of Europe, a great part of Asia, the north of Africa, and some parts of North America. Having this wide range, and growing in situations where it could not fail to be conspicuous, it is mentioned by the earliest poets and writers on natural history ; thus Homer :

> " From out the cover'd rock,
> In living rills a gushing fountain broke;
> Around it and above, for ever green,
> The bushy Alders form'd a shady scene."

The poet Virgil assigns to the Alder the distinction of being the tree out of which the first boat was shaped :

> " Tunc Alnos primum fluvii sensere cavatos."
> " Then rivers first the hollowed Alder knew."
>
> " Nec non et torrentem undam levis innatat Alnus."
> " And the light Alder skimm'd the torrent wave."

Theophrastus was acquainted with its property of dyeing leather, and Vitruvius recommends the wood for piles, stating that the city of Ravenna was built on it.

The Alder, in its young state, is a bushy shrub of a pyramidal form, heavily clothed with large,

deep green leaves, which as well as the young
shoots are covered with a glutinous substance,
more especially in the early part of the summer.
The leaves are roundish, blunt and serrated,*
shining above, and furnished at the angles of the
veins beneath with minute tufts of whitish down.
The leaf-stalks are nearly an inch in length, and
furnished with stipules,† which entirely enclose
the leaves before their expansion. The flowers
are of two kinds; the barren are long drooping
catkins which appear in the autumn, and hang on
the tree all the winter, and the fertile are oval,
like little Fir-cones, but are not produced until
spring. When these ripen, the thick scales of
which they are composed separate, and allow the
seeds to fall, but remain attached to the tree
themselves all the winter, and by them the tree
may be distinguished when stripped of all its
foliage. In young trees the bark is smooth and
of a dark purple-brown hue, but in old trees it is
rugged and nearly black. When allowed to attain
its full growth, it reaches a height of forty or
fifty feet, if the situation be favourable; but in
the mountains and in high latitudes it does not rise
above a shrub. Wordsworth, in one of his son-
nets to the river Duddon, says:—

"But now to form a shade
For thee, green Alders have together wound
Their foliage; Ashes flung their arms around,
And Birch-trees risen in silver colonnade."

There are probably few rivers in England
which have not Alders growing somewhere

* Serrated, notched like a saw.   † Stipules; see vol. i. p. xix.

or other on their banks. Where they most
flourish is in good soil, which is at all times a
little raised above the level of the water; for
although they will grow in swampy ground, they
prefer places where their roots are not always
covered with water. It has been observed that
their shade is much less injurious to vegetation
than that of other trees:

> The Alder, whose fat shadow nourisheth—
> Each plant set neere to him long flourisheth.
> BROWNE.

The haunts of the Alder being the places where
beyond all others we should expect to find picturesque scenery, it cannot fail to form a part of
many a beautiful landscape, though it contributes
but little itself, the outline of the tree being
in most cases too formal, and the foliage not
broken into varied masses. Yet it has ·its admirers. Gilpin considers it, "the most picturesque of the aquatic trees, except the weeping
willow. He who would see the Alder in perfection must follow the banks of the Mole, in Surrey,
through the sweet vales of Dorking and Mitcham
into the groves of Esher. The Mole, indeed, is
far from being a beautiful river; it is a silent and
sluggish stream; but what beauty it has it owes
greatly to the Alder, which everywhere fringes
its meadows, and in many places forms very
pleasing scenes." Sir T. D. Lauder is of the
same opinion: "It is," he says, "always associated in our minds with river scenery, both of
that tranquil description most frequently to be
met with in the vales of England, and with that
of a wilder and more stirring cast, which is to be

found amidst the deep glens and ravines of Scotland. In very many instances we have seen it put on so much of the bold, resolute character of the Oak, that it might have been mistaken for that tree, but for the intense depth of its green hue. Nowhere will the tree be found in greater perfection than on the wild banks of the river Findholm and its tributary streams, where scenery of the most romantic description everywhere prevails." Trees of similar character are not uncommon on the banks of rivers in other parts of Scotland, and in the north of England. On the whole, though the Alder does not take a high rank among our picturesque trees, we must recollect that it often flourishes where no other tree would live, and thus ornaments a landscape which would otherwise be tame and naked. It retains its leaves, too, until very late in the year, and gloomy though their tone may be, we forget this defect when nearly all other trees are bare.

The principal use of the Alder, when growing, is to prevent the encroachment of rivers on their banks, especially where a stream flowing through a loose soil makes a sudden turn. Planters do not recommend its being planted to fill up places in moist woods, but that such ground should be drained and planted with other trees. "For such is the nature of the Alder that it attracts and retains the moisture around it. This effect is occasioned by the nature of its roots, which are chiefly composed of a huge mass of small fibres, whose capillary attraction prevents the escape of a redundant water in the vicinity of the plants. This property of creating swamps we have re-

peatedly observed in the Alder, and, from experiments we have made, are fully convinced that
a plantation of Alders would soon render the
ground, (even should it be previously of tolerably
sound and dry quality) soft and spongy, and in
time convert it into a decided bog."*

The wood of the Alder is soft and light, and if
exposed alternately to wet and dry, will scarcely
last a year; but if kept entirely submersed or
buried in damp earth, no wood is more durable.
Hence it is extensively used for foundations of
bridges, water-pipes, pumps, &c.

By lying for a long time in peat bogs, it acquires a black hue, but from its softness will not
take a good polish. The young branches are
much used for the purpose of filling in drains,
and are more durable than any other kind of
brushwood. Sir T. D. Lauder says that the
wood is very valuable, even when of a small size,
for cutting up into herring-band staves. Old
trees which are full of knots, may be made into
tables, and chairs which, if protected from insects
by French polish, are both beautiful and durable.
The charcoal is highly valued in the manufacture
of gunpowder, for which purpose it is in some
places largely planted. The colour of the wood
when first cut is white, it soon, however, becomes
of a bright red, which afterwards fades into pink,
which is its permanent hue. Few river-side wanderers have failed to notice the bright tints of the
chips and newly hacked trunks which have here
and there marked the recent labours of the woods-
man. The bark and young shoots are used for
tanning as well as for dyeing several tints; com-

* Selby.

bined with iron it produces a very good black.
The Alder is increased either by seeds or by
truncheons, the latter method being preferred in
places which are liable to be overflowed, and

CUT-LEAVED ALDER.

where consequently a firm hold in the ground is
desirable. Several varieties are cultivated with
leaves cut like those of the Hawthorn and Oak,
and these frequently attain a large size.

Fine specimens of the Alder grow in the Bishop
of Durham's Park at Bishop Auckland; but that
which is said to be the largest in England stands
near a rivulet at Haverland in Norfolk. It is
sixty-two feet high; and the trunk, at one foot
from the ground, measures eleven feet seven inches
in circumference. Of this tree an excellent figure
is given in Grigor's " Eastern Arboretum."

II.   H

THE ELM.

THE ELM.

ULMUS.

Natural Order—ULMACEÆ.

Class—PENTANDRIA.   *Order*—DIGYNIA.

OF this tree, to which the cultivated parts of England are so much indebted for the richness of their landscape, there are many varieties. No less than eighteen are described by Loudon which are all referred to the commonest species, *Ulmus campestris.* It will not be necessary here to supply even a catalogue of these, and it would be impossible to point out the distinctive characters of each without entering into a tedious and unprofitable description, which the reader, if he wishes to study the Elms botanically, will be able to find in other works. Botanists are far from being agreed as to which should be termed species, and which varieties, so uncertain are the characters; nor shall I attempt any settlement of the question, but, omitting all notice of the rarer and less strongly marked kinds, mention those only which are either universally allowed to constitute species, or, at least, very distinct varieties.

All the Elms indigenous to Great Britain or naturalized, agree in the following characters:— They are lofty trees, having a straight columnar trunk, with hard wood, a rugged bark, and zigzag, slender branches, which, when young, are either downy or corky. In winter and early

spring many of the young twigs may be observed
thickly set with bead-like scaly buds, which expand
before the leaves, and contain the rudiments of
flowers, each of which consists of a calyx of one

BRANCH OF ELM.

leaf divided into five purple segments, and enclosing an equal number of stamens of the same
colour, and a cloven germen bearing two styles.*
The stamens soon wither and fall off, but the germen enlarges and becomes a thin, pale, membranous seed-vessel, rounded and notched at the
extremity, and bearing in the centre a solitary
seed. The calyx remains attached to the base of

* For an explanation of these terms, see vol. i. p. 21—23.

the seed-vessel, which does not open, but serves
as a wing to waft away the ripe seed, if it does
ripen, which is not always the case. So numerous and conspicuous are these seed-vessels, that
they might be mistaken, as indeed they sometimes
are, when seen from a distance, for tufted foliage,
an error which is all the more likely to occur

SEED-VESSEL.

because the leaves rarely begin to expand until
the seeds are nearly ripe. Few persons can have
failed to notice the numerous leaf-like plates flut-
tering tremulously through the air, during the
high winds of April, or sweeping in eddies along
the road in the neighbourhood of Elm-rows.
These are the seed-vessels just described; and
there is something melancholy in the sight of them,
reminding us, as they do, of autumn and the fall
of the leaf, before spring has well set in. Towards

the end of April the leaves burst forth from another set of buds : they are at first of a fresh,
bright green, but afterwards deeper in tint, irregularly notched at the edge, and remarkably unequal at the base, more or less rough on both
sides, prominently veined beneath, and having a
downy tuft where each vein commences. Each
leaf has a pair of oblong stipules, which, however,
soon fall off.

Thus far the description given will apply to all
the species of Elm; we will now proceed to consider the leading characters which distinguish the
four commonest species.

Ulmus campestris, Common Small-leaved Elm.

FLOWERS OF COMMON ELM.

This is the most generally distributed species of

all. It is a lofty, upright tree, composed of many tiers of spreading branches, which often hang in graceful festoons at the extremities; its flowers are not easily distinguished from those of other

COMMON ELM.

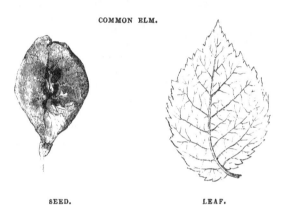

SEED.   LEAF.

species; the winged seed-vessel is cleft nearly down to the centre; the leaves are rough to the touch, tapering to a point, and obliquely wedge-shaped at the base; the young twigs are downy, and sometimes slightly corky.

Ulmus stricta, Cornish Elm, is a tree of more rigid growth than the preceding; the flower-buds are arranged more regularly on the twigs; the leaves are much smaller, more evenly notched, and nearly smooth. It is confined to the counties of Devon and Cornwall. In these two species the main trunk is generally continuous nearly to the summit.

Loudon says of this variety (as he calls it), "that in the climate of London it is a week or fortnight later in coming into leaf than the Common Elm.

CORNISH ELM.

It attains a very great height, and has a somewhat narrower head than the other kinds. This is also the character of the tree in the west of England; but as it is generally grown in hedges, where frequent loppings prevent it from assuming its natural shape, it is by no means a picturesque tree. The timber is said by many to be superior to that of any of the other Elms.

Lindley describes, under the name of *Ulmus parvifolia*, a variety with much smaller leaves; but this appears to be little known. Loudon mentions also another Cornish variety of Elm, which is almost evergreen in a mild winter; and as such is the most ornamental tree of the genus. It is called the Kidbrook Elm.

The Cornish Elm cannot be considered a picturesque tree. It is of a rigid growth; the foliage is meagre and rarely hangs in graceful clusters. The timber, however, is considered very good.

Ulmus montana, Wych Elm, is well distinguished from the preceding by its numerous spreading branches, which frequently droop so as to conceal

FLOWERS AND SEED-VESSELS OF WYCH ELM.

the main trunk; its flowers are in looser tufts than those of the Common Elm; and the seed-vessel differs materially in being only slightly notched, instead of cleft to the centre; the leaves are much larger, sharp-pointed, and nearly equal at the base.

Ulmus suberosa, Cork-barked Elm, is in habit between the Common and Wych Elms, being more spreading than the former, but not so much so as the latter. The leaves are very large; but the best distinctive characters are afforded by the branches, which, when one year old, are very

BRANCH OF ULMUS SUBEROSA.

hairy, and in the second year are thickly coated with a cracked, corky, excrescence, from which the tree derives its name. A foreign species, called Dutch Elm, has also corky branches, but the young twigs are always smooth.

Two other British Elms are described by botanists, but, as they are of local occurrence, a notice of them will not interest the general reader. The above characters, it is hoped, will be sufficient to enable the student to determine the species of any Elm which he is likely to meet with.

The Elm was well-known both to the ancient Greeks and Romans; the former were acquainted with two species which grew severally in the

mountains and the plains. Pliny enumerates four
species which were known to the Romans, the
Atinian (the same as our common small-leaved
Elm), the Gallic, the Italian, which had tufted
foliage, and the Wild Elm. These appear to
have been equally valued for their leaves, which
were given as fodder to cattle. The Gallic and
Italian kinds were preferred to every other tree
as a support to vines, for which purpose they were
planted in regular rows at set distances, such plantations being called "arbusta." The rearing of
the trees was considered of such importance, that
Pliny gives specific directions for the formation of
an *Ulmarium*, or plantation of Elms, directing that
the seeds should be gathered in March, sown in
beds, and the young trees planted out in nursing
beds before they took their station in the vineyard.
He directs also, that when transplanted, it should
always be to a similar or better soil, and even
recommends that the bark should be marked while
they stood in the nursery, in order that when
transplanted where they were to remain, their
northern sides might retain the same aspect. If
reared from suckers, he directs that they should
be planted out in autumn. The Atinian Elm was
never used as a vine-prop, on account of its too
luxuriant foliage, which kept off the sun from the
ripening grapes. An important part of the vine-
dresser's occupation was to prune the Elms, which,
when the vine was trained to them, were said to
be "married." The minuteness of these directions, which are also alluded to by the Roman
poet Virgil, proves the estimation in which the
tree was held; and the name of the fourth species, "wild," would seem to shew that that species

was not considered to be so well adapted to the purpose as the cultivated kinds. The leaves and bark were supposed to have an astringent property, and were therefore used in the curing of wounds; the timber was recommended, for its rigidity and toughness, as fit for the hinges, or rather pivots, of gates; and Virgil tells us, that young Elms were bent down while in a growing state, and kept in a curved position until they had acquired the necessary shape, in order to be converted into plough-tails, a process which has been imitated in modern times with respect to Oak trees, for the production of what is called knee-timber in ship-building. The wedding of the Vine to the Elm is frequently mentioned by the Roman poets among the tranquil and healthful occupations of rural life. Some authors are of opinion that the Elm was introduced into Britain by the Romans along with the Vine, and this opinion borrows weight from the fact, that it rarely matures its seeds, and therefore would require the assistance of man to secure its continued propagation. Since, too, the Elm was one of the trees frequently planted on funeral mounds, it may have been introduced for that purpose, while the similarity of the English name, *Elm*, to the Latin *Ulmus*, seems to confirm the opinion of the foreign origin of the tree.

Evelyn sagely remarks: " It seems to be so much more addicted to some places than to others, that I have frequently doubted whether it be a pure indigene or translatitious (introduced); and not only because I have hardly ever known any considerable woods of them, but almost continually in tufts, hedge-rows and mounds; and that Shrop-

shire, and several other counties, have rarely any
growing in many miles together. In the mean-
time, some affirm they were first brought out of
Lombardy, where indeed I have observed very
goodly trees about the rich grounds, with Pines
among them." Dr. Hunter, however, Evelyn's
editor, is of opinion that " the Elm is certainly a
native of this country;" and he has much reason
on his side, for the Atinian Elm, which is universally considered to be the same with our common Elm, did not, according to Pliny, ripen its
seeds in Italy, any more than it does in England.
But in this country, as well as in that, it produces
abundance of suckers, and it is by no means uncommon for plants that increase freely by roots to
produce abortive seed-vessels. The Great White
Convolvulus* or Bindweed, for instance, and the
Lesser Periwinkle,† which are most prolific by
their roots, and are undoubted natives, have never
been known to perfect their seeds. Besides which,
the authors who maintain that the Elm was introduced into Britain as a companion of the Vine,
appear to have lost sight of Pliny's assertion that
the Atinian, or Common Elm, was never used for
the purpose, on account of its excessive foliage.‡
On the whole, then, the Elm has as good a claim
to be considered a native of Britain, as of any of
the other European states, not excepting even
Italy, from which it is said to have been brought.

Gilpin, speaking of the Elm as a picturesque
tree, has the following remarks : " The Oak and
the Ash have each a distinct character. The

---

* Convolvulus sepium. † Vinca minor.
‡ Prima omnium ulmus, exceptâ propter nimiam frondem Atiniâ.—
Plin. Sec. Nat. Hist. lib. xvii., *cap.* xxiii.

massy form of the one, dividing into abrupt, twisting, irregular limbs, yet compact in its foliage,
and the easy sweep of the other, the simplicity of
its branches, and the looseness of its hanging
leaves, characterize both these trees with so much
precision, that at any distance at which the eye
can distinguish the form, it may also distinguish
the difference.   The Elm has not so distinct a
character.   It partakes so much of the Oak, that
when it is rough and old, it may easily, at a little
distance, be mistaken for one; though the Oak,
I mean such an Oak as is strongly marked with
its peculiar character, can never be mistaken for
the Elm.   This is certainly a defect in the Elm;
for strong characters are a great source of picturesque beauty.   This defect, however, appears
chiefly in the skeleton of the Elm.   In full foliage
its character is better marked.   No tree is better
adapted to receive grand masses of light.   In this
respect it is superior both to the Oak and the
Ash.   Nor is its foliage, shadowy as it is, of the
heavy kind.   Its leaves are small, and this gives
it a natural lightness; it commonly hangs loosely,
and is in general very picturesque.   The Elm
naturally grows upright; and when it meets
with a soil it loves, rises higher than the ge-
nerality of trees; and after it has assumed the
dignity and hoary roughness of age, few of its
forest brethren (though, properly speaking, it is
not a forester) excel it in grandeur and beauty.
The Elm is the first tree that salutes the early
spring with its light and cheerful green,—a tint
which contrasts agreeably with the Oak, whose
early leaf has generally more of the olive cast.
We see them sometimes in fine harmony together,

about the end of April and the beginning of May. We often, also, see the Elm planted with the Scotch Fir. In the spring, its light green is very discordant with the gloomy hue of its companion; but as the year advances, the Elm leaf takes a darker tint, and unites in harmony with the Fir. In autumn, also, the yellow leaf of the Elm mixes as kindly with the orange of the Beech, the ochre of the Oak, and many of the other fading hues of the wood."

This is, undoubtedly, high praise of the Elm, higher perhaps than the author intended; for the principal faults which he finds with the tree can be detected only in the winter, the season when we look anywhere rather than to the woods for picturesque beauty. If, moreover, we consider the frequent occurrence of the Elm in situations where, without it, the landscape would be a blank, but with it, is exceedingly picturesque; and when we add to Gilpin's list of recommendations the very important one which he has omitted, namely, the graceful outlines of its festoons of foliage, painted on an evening sky, we cannot but allow that it ranks very high among the natural ornaments of an English landscape.

Frequent mention of the Elm occurs in the English poets, who also allude to the Italian custom mentioned above. Spencer calls it the " Vineprop Elm :" Milton, describing the daily occupations of Adam and Eve in Paradise, says :—

> " They led the Vine
> To wed her Elm ; she, spoused, about him twines
> Her marriageable arms, and with her brings
> Her dower, the adopted clusters, to adorn
> His barren leaves."

And Wordsworth, having, probably, the same passage in his mind, says :—

> " Our days glide on ;
> And let him grieve, who cannot choose but grieve,
> That he hath been an Elm without his Vine
> And her bright dower of clustering charities,
> That, round his trunk and branches, might have clung
> Enriching and adorning."

With Milton it was a great favourite, from its affording a shady retreat to the care-worn citizen :—

> " Not always city pent, or pent at home,
> I dwell ; but when spring calls me forth to roam,
> Expatiate in our proud suburban shades
> Of branching Elm, that never sun pervades."

The " rugged Elm " is enshrined too in Gray's Elegy, as every one will recollect.

> " Beneath those rugged Elms, that Yew-tree's shade,
> Where heaves the turf in many a mould'ring heap,
> Each in his narrow cell for ever laid,
> The rude forefathers of the hamlet sleep."

The Elms immortalized by the poet have been cut down ; but the Yew-tree is still in existence.

The Common Elm is generally propagated by suckers, which spring up in great abundance round the trunk, or by grafting on young plants of Wych Elm which have been raised from seed. It grows most rapidly in light land, but requires a stiff strong soil to produce good timber. It will bear any amount of pruning, but needs none : the custom of lopping Elms in hedge-rows, and converting them into gigantic brooms, is as injurious to the timber as it is destructive of picturesque effect. It is to be presumed, however, that farmers who adopt this practice are remunerated by the additional produce of their lands, thus thrown open to the sun and air. The Elm bears transplanting remarkably well even at an advanced age ; hence it is well adapted for planting in the neighbourhood of modern houses, where a speedy shade is desired. For avenues it is unrivalled, forming a delightful shade, and crossing at a lofty elevation and at the exact angle which is most pleasing to the eye. The avenue of these trees at Strathfieldsaye, the seat of the Duke of Wellington, is a mile in length, and is greatly admired.

In ancient times the leaves of the Elm were much used as fodder for cattle, and this is still the case in many parts of the Continent. Evelyn recommends the revival of this practice in England in seasons when the hay-harvest is defective : he states, that cattle prefer them to oats, and thrive exceedingly well on them : the inner bark is very tough, and, like that of the Lime, is made into bast mats and ropes ; the timber is fine-grained and tough, and is remarkable for its durability under water. Hence it is highly prized in naval architecture, being used for the keels of large vessels, and many parts of the rigging which are most liable to exposure to wet. It was formerly also much used for making water-pipes, but has, within the last few years, been almost superseded for this purpose by cast-iron pipes.

The Elm, growing in a forest, and in good soil, arrives at perfection in a hundred and fifty years, but it will live for five or even six hundred years. Large forest Elms are cut down with advantage when of an age between 100 and 130 years, and then furnish a large quantity of building material.

Elms which have been lopped live for a shorter
period than others, and should therefore be cut
down when no more than seventy or eighty years
old.

A small-leaved species of Elm is selected by the
Chinese to be treated in the way described in
Vol. i. page 306, as being adopted with regard to
the apple. A young tree is planted in a pot, and
surrounded with pieces of rough stone to represent rocks, among which mosses and lichens are
introduced. It is not suffered to rise higher than
about a foot or fifteen inches. No greater supply
of water is given than just enough to keep it alive,
and every means is used to give it a stunted appearance. The points of the shoots, and the half
of every new leaf, are constantly and carefully cut
off; the stem and branches are distorted by means
of wires, and the bark is lacerated to produce a
rugged character. One branch is partly broken
through, and allowed to hang down as if by accident; another is mutilated to represent a dead
stump. This treatment produces, in course of
time, the appearance of an old weather-beaten
forest-tree, and it is then, if unworthy of all the
pains that have been bestowed upon it, certainly a
curious object.

Several insects prey on the Elm, among which
by far the most mischievous is the Elm-destroying
Beetle (*Scolytus destructor*). The ravages committed by this minute insect, would scarcely be
credible, were we not informed that as many as
80,000 have been found in a single tree. Two
eminent entomologists, Mr. Spence in England
and M. Audouin in France, have turned their
attention to this subject, and have satisfactorily

shewn the importance of watching the habits of an
insect less than a quarter of an inch in length.
The result of their observation is that the perfect
insect feeds on the inner bark of the Elm, to reach
which it perforates the outer bark, and feasts at its
leisure.   The cavities thus made interrupt the ascent and descent of the sap, and retain moisture,
from the combined effect of which causes the tree,
in the course of a few years, becomes sickly, and
is brought into exactly that state in which the fe-
male selects it for laying her eggs; though sometimes she attacks a tree which is beginning to
decay from other causes.   A suitable tree having
been selected about July, she perforates the bark,
and eats away a vertical passage about two inches
in length, laying from twenty to fifty eggs as she
advances.   Having
completed her task,
she dies.   About two
months afterwards the
eggs are hatched, and
the grubs immediately
begin to eat their way
also through the inner
bark in a horizontal
direction, some to the
right and some to the
left, but never inter-
fering with each other's
track.   When   each
grub has finished its
course of eating, it
turns to a pupa and
then to a beetle; after which it gnaws a straight
hole through the bark, and comes out about

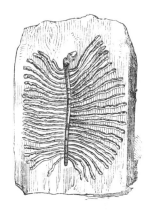

ELM-DESTROYING BEETLE.

the end of May, in the year following that in which the eggs were deposited. The injury thus inflicted by the grubs is so much greater than that occasioned by the perfect insect, that, when the former have commenced their ravages, the tree cannot be saved from destruction, and the only alternative is to cut it down and to burn every particle of bark. But when a tree is attacked by the perfect insect, it may be saved by being carefully brushed over with coal-tar, the smell of which is so offensive that the insects will desert it, and in the course of a few years it may recover its healthy condition. Had this discovery been made and acted upon at an earlier period, it is probable that an immense number of trees in the promenades of several of the principal cities of Europe, including from twenty to thirty in St. James's Park, might have been saved.

In the second volume of Kerby and Spence's "Entomology," p. 478—480, is given the history of a minute moth which feeds on the leaves of the Elm, and in the construction of its moveable dwelling displays an instinct closely approaching reason. The passage is too long to quote, but it will be found well worthy of perusal, as shewing that the insect is not guided by a blind undeviating instinct, but varies its operations according to the materials which it has within reach, at one time sewing its mantle together with silk, at another availing itself of the natural union between the upper and under surface of the leaf.

The Elm retains its foliage till late in the autumn, the leaves assuming a rich yellow hue some time before they fall from the tree. If examined closely at the season, they will be found to be marked with dark-coloured blotches. These spots contain the instruments appointed for ensuring the decay of the leaves. During the winter months the leaves remain on the ground unaltered except

SPOTTED ELM-LEAF.

in colour; but in spring the spots become matured, the surface cracks, and a minute fungus appears: decomposition spreads from these points, and the leaves very soon decay.

Among the most remarkable Elm-trees recorded by various authors as now existing in England, the following are most worthy of mention.

The Chipstead Elm, an admirable engraving of which is given by Strutt in his " Sylva Britannica," stands in the park at Chipstead Place, in Kent. It is sixty feet high; twenty feet in circumference at the base; and fifteen feet eight inches, at three feet and a half from the ground. It contains 268 feet of solid timber; but this bulk is comparatively small compared to what it would have been, had it not sustained the loss of some

large branches near the centre. Its venerable
trunk is richly mantled with ivy, and gives signs of
considerable age; but the luxuriance of its foliage
attests its vigour, and it is as fine a specimen of
its species in full beauty as can be found.

The Crawley Elm, described by the same author, is a well-known object to travellers by the
high-road between London and Brighton, attract-
ing attention by its tall and straight stem, which
is seventy feet high, and by the fantastic rugged-
ness of its widely-spreading roots. Its trunk is
perforated to the very top, and measures sixty-
one feet in circumference at the ground, and
thirty-five feet round the *inside*, at two feet from
the base. There is a regular door to the cavity in
this tree, the key of which is kept by the lord of
the manor; but it is opened on particular occasions, when the neighbours meet to regale themselves within the cavity, which is capable of containing a party of more than a dozen. The floor
is paved with bricks. Madame de Genlis says
that a poor woman gave birth to a child in the
hollow of this tree, and afterwards resided there
for a long time.

Scarcely less remarkable than this, was the
Northover Elm, which a few years ago stood on the
lawn of Northover House, Ilchester. It measured
fifty-eight feet nine inches in girth close to the
ground, and fifty-four feet six at a height of ten
feet. At about thirteen feet from the ground
it threw out seven large branches in a circular
manner. On these was constructed a room, capable of containing twenty persons, in which the
late Mr. Chichester, one of the magistrates of
the county of Somerset, was, during the summer

season, accustomed to hold his justice meetings. This tree was blown down on the 23rd of December 1833.

The famous " Gospel Elm," which formerly stood at Stratford on Avon, and was held in veneration, not only for its size, but for the pious custom which its name commemorated,* was cut down in 1847, an impression having gained ground that it was hollow and consequently unsafe. When, however, the trunk was sawn through, it was found, too late, to be perfectly sound, and the grain of the tree beautifully marked, particularly towards the crown; scarcely a blemish or flaw was discernible throughout, which materially tended to increase the very general regret expressed that so interesting a relic should have been sacrificed. It was sold by auction, and realized the sum of 23*l.* 13*s.*, being purchased to be manufactured into pieces of furniture.

Mr. Jesse mentions in his " Gleanings" a tree which possesses considerable historical interest. " It is, perhaps," he says, " not generally known that one of the Elm-trees standing near the entrance of the passage leading into Spring Gardens, was planted by the Duke of Gloucester, brother to Charles I. As that unfortunate monarch was walking with his guards from St. James's to Whitehall, on the morning of his execution, he turned to one of his attendants and mentioned the circumstance, at the same time pointing out the tree."

* See vol. I. p. 91.

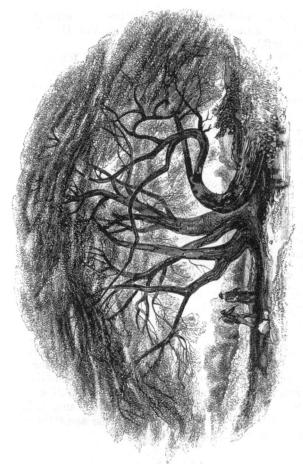

WYCH ELM AT ENYS, CORNWALL.

THE WYCH ELM.

THIS species, it has been observed above, may be distinguished from the Common Elm by its larger leaves and slightly-notched seed-vessels. A practised eye will also be able to detect it readily by other peculiarities. The shoots of the young trees are of so vigorous a growth as to be nearly equal in size to the stem from which they spring; they are also so heavily laden with leaves, which are as large as those of the Hazel, or even larger, that they have an arched, drooping appearance. On the older branches the leaves are smaller, and hang in large heavy masses; they may be distinguished by being taper-pointed, and nearly equal at the base. The trunk is less upright than those of the other species, and soon divides into long, widely-spreading, somewhat drooping branches. Though less common in England than the Small-leaved Elm, it is far from rare. In Scotland, it is the only indigenous species, whence it is often called the Scotch Elm. From the leaves somewhat resembling those of the Hazel, Gerard tells us it is sometimes called the "Witch-hasell." "Old men affirm," he adds, "that when long bows were in use, there were very many made of the wood of this tree; for which purpose, it is mentioned in the English statutes by this name of Witch-hasell." The meaning of its name is unknown; hence it is variously spelt by authors, wych, wich, witch, and

weech.   In some of the midland counties, the
name seems to have originated the notion that it
is a preservative against witchcraft, and a sprig
is inserted into a hole in the churn by dairy-maids,

LEAF OF WYCH ELM.

in order that the butter may come freely.   The
foliage withers much earlier than that of the
Common Elm, curling up and becoming brown
before almost any other tree has acquired its autumnal tint.

Gilpin considers the Wych Elm " generally
more picturesque than the common sort, as it
hangs more negligently; though, at the same time,

with this negligence, it loses in a good degree that happy surface for catching masses of light which we admire in the Common Elm. We observe, also, when we see this tree in company with the Common Elm, that its bark is somewhat of a lighter hue." Sir Thomas D. Lauder is of the same opinion. " The trunk is so bold and picturesque in form, covered, as it frequently is, with huge excrescences; the limbs and branches are so free and graceful in their growth, and the foliage is so rich, without being heavy or clumsy as a whole, and the head is generally so finely massed, and yet so well broken, as to render it one of the noblest of Park-trees ; and when it grows wildly amid the rocky scenery of its native Scotland, there is no tree which assumes so great or so pleasing a variety of character.

The Wych Elm ripens its seeds freely in June, but produces no suckers ; it grows more rapidly than the common kind : and this probably is the reason why its timber is inferior for most purposes. It is nevertheless valuable to the wheelwright and millwright, and the excrescences are highly prized by the cabinet-maker, who makes of them a beautiful veneer for tables, work-boxes, &c. The bark of the young limbs is very tough and flexible, and is often stripped off in long ribands, and used, especially in Wales, for securing thatch and for other similar purposes.

Though the Wych Elm does not produce suckers, it strikes from layers with great facility, and if a growing branch or twig by any accident touches the ground, it is sure to take root. A striking instance of this is afforded by a tree at Enys, in Cornwall, of which an engraving is given at page 120. It was planted originally on the left side of a little stream, but having, from some unknown cause, been laid prostrate, the trunk fell on the opposite side of the stream, where it took root, and rising again, has acquired such dimensions that it covers an area of seven thousand square feet, or one sixth of an acre. The main stem, which now forms a natural bridge across the stream, is ten feet three inches in circumference, and the three trunks, which rise from the right side of the stream, measure severally, eight feet and a half, six feet, and five feet eight inches.

These trees sometimes attain a great size. Evelyn mentions one " growing within these three or four years in Sir Walter Baggot's park, in the county of Stafford, which, after two men had been five days felling it, lay forty yards in length, and was at the stem seventeen feet diameter; it broke in the fall fourteen load of wood, forty-eight in the top; yielded eight pair of naves, eight thousand six hundred and sixty feet of boards and planks. It cost ten pounds seventeen shillings the sawing; the whole esteemed ninety-seven tons. This was certainly a goodly stick!"

The Cork-barked Elm resembles the Common Elm rather than the Wych Elm; it rarely ripens its seeds, but produces suckers freely. The timber is soft and spongy, and much inferior to that of either of the others.

THE HORNBEAM.

THE HORNBEAM.

CARPINUS BETULUS.

Natural Order—AMENTACEÆ.

Class—MONŒCIA.   *Order*—POLYANDRIA.

OF all our indigenous forest-trees perhaps no one is so little known as the Hornbeam; nor is this surprising, for, although it frequently reaches a height of fifty or sixty feet, it has no strongly-marked distinctive character, and is often mistaken for some kind of Elm, to which its foliage bears a great resemblance. It is found in most of the temperate countries of Europe and Asia, and is far from uncommon in several of the counties of England; in some it is so abundant that it forms (as Sir J. Smith observes) a principal part of the ancient forests on the north and east sides of London: such as Epping, Finchley, &c. By the Greeks it was called *Zugia*, or "yoke-tree," from the use to which its timber was applied; the Latins called it *Carpinus*, the name by which it is still known to botanists.

It has a straight and tolerably smooth trunk, which is slender and very frequently flattened, twisted, or otherwise irregular in shape, and is subdivided into a large number of long tapering branches, which diverge in such a way that the main stem is generally lost in the confused mass at some distance below the summit. The branches are remarkably liable to unite when they touch in

crossing, hence very curious appearances are sometimes produced. The outline of the head is round,
and possesses little picturesque beauty. The leaves
are shaped somewhat like those of the Beech,
but are rough and notched at the edge like those

LEAF OF THE HORNBEAM.

of the Elm; they may be distinguished from the
former by their roughness, and from the latter by
their being plaited when young, and by having
numerous, regular, strongly marked veins.   Like

the Beech, too, they retain their withered foliage
on the young branches all the winter. The Hornbeam when young is also very similar in habit to
the Beech, but the latter may immediately be
detected, on examination, by its glossy leaves.
The flowers appear soon after the leaves, in April,
growing in catkins of two kinds, of which
the fertile are succeeded by clusters of
small angular nuts, each seated at the
bottom of a leafy cup. When these are once
formed, the tree which bears them cannot be mistaken, for no other British tree bears fruit of the
same kind. The leaf buds are longer and sharper
than those of the Elm.

Owing to its partaking several of the properties of other trees, some of the old writers were
puzzled to find its place in the system. Pliny
probably saw some resemblance between its clusters of nuts and the keys of the Maple, for he
places it among the ten kinds of Maple, but adds,
that others considered it to belong to a distinct
genus. Its second name, Betulus, would seem to
imply that it was, by some of the early botanists,
considered a kind of Birch, and one of its old
English names, " Witch-hasell," points to the supposition that it was a kind of Hazel. Gerard says,
" It growes great and very like unto the elme or
wich hasell tree; having a great body, the wood
or timber whereof is better for arrowes and shafts,
pulleyes for mils, and such like devices, than elme
or wich hasell; for, in time, it waxeth so hard,
that the toughnes and hardnes of it may be
rather compared unto horn than unto wood; and
therefore it was called hornebeam or hard-beam.
The leaves of it are like the elme, saving that they

II. K

be tenderer: among these hang certain triangled things, upon which are found knaps, or little buds of the bignesses of ciches (vetches), in which is contained the fruit or seed. The root is strong and thicke."

Evelyn is loud in his praises of the Hornbeam; for, the tree being, as it is called, "tonsile," or very patient of being clipped by the shears, it was highly prized in the formal gardens of his day. "It makes," he says, "the noblest and stateliest hedges for long walks in the gardens or parks, of any tree whatsoever whose leaves are deciduous and forsake their branches in winter, because it grows tall and so sturdy as not to be wronged by the winds; besides, it will furnish to the very foot of the stem, and flourishes with a glossy and po-lished verdure, which is exceedingly delightful, of long continuance, and, of all the other harder woods, the speediest grower, maintaining a slender upright stem, which does not come to be bare and sticky in many years. It has yet this (shall I call it) infirmity, that, keeping on its leaves till new ones thrust them off, it is clad in russet all the winter long. That admirable espalier hedge, in the long middle walk of the Luxemburgh garden, at Paris, than which there is nothing more grace-ful, is planted of this tree; and so is that cradle or close walk, with that perplext canopy which lately covered the seat in his Majesty's garden at Hamp-ton Court. They very frequently plant a clump of these trees before the entries of the great towns in Germany, to which they apply timber-frames for convenience of the people to sit and solace in."

Dr. Hunter tells us, that the Hornbeam was in

great repute at the close of the last century for hedges. The plants were raised from layers, and set in single rows in a sloping direction, so that they crossed one another like large net-work. The parts where the stems crossed were stripped of their bark and bound together with straw. By this process they united into a firm palisade, and throwing out numerous shoots, in a few years formed an impenetrable fence. It was not uncommon, he says, to see the sides of high roads thus guarded for many miles together.

The taste for forming "labyrinths," "stars," "alcoves," and "arcades," happily having now passed away, the Hornbeam is only admitted into gardens for the purpose of forming hedges to shelter tender plants, and for this its numerous branches and the property which it possesses of retaining its leaves during winter well adapt it. Another recommendation is, that it grows well in the coldest and hardest soils, and may consequently be employed where other trees would not thrive.

The wood of the Hornbeam is white and close-grained, and though not flexible, surpasses in toughness the timber of any other British tree. The unevenness of the trunk described above is, however, communicated to the fibre of the wood, and hence it does not take a good polish. This defect does not exist in the young wood, which is exceedingly well adapted for the yokes of cattle and all kinds of wheelwright's work, especially mill-cogs. Selby recommends that it should be planted extensively in cold, stiff, clayey soils, for the staves of fish-barrels. It ranks among the best of fuels, burning freely, and giving out a great deal of heat; it is highly inflammable, lighting easily and making a bright flame. This property was known to the ancients, for Pliny speaks of its being used for marriage torches. Its charcoal is highly prized, not only for ordinary purposes, but for the manufacture of gunpowder. The inner bark is also used, according to Linnæus, for dyeing yellow.

A number of trees are recorded by Loudon averaging from fifty to eighty feet high, with trunks from six to nine feet in circumference, but none requiring any particular notice. At Aldermaston Park, in Berkshire, is a group of fine Hornbeams, which were evidently planted to form one of the quaint devices so much in vogue in the seventeenth century. They surround an elliptical area thirty paces in length and fifteen in width, and, crossing their branches high over head, form a leafy dome far more imposing than anything which the planter could have contemplated. The original intention, probably was, that they should have been trained to form a hedge, such as Evelyn loved to look upon; but they have long escaped from this unnatural thraldom, and now rise to a height of fifty or sixty feet, with trunks varying from three to seven feet in circumference, and beautifully covered with lichens.

The Hop Hornbeam, occasionally met with in gardens and pleasure-grounds, approaches the common Hornbeam in character, but belongs to the genus *Ostrya*. It is not a native of Britain.

THE HAZEL.

CORYLUS AVELLANA.

Natural Order—AMENTACEÆ.

Class—MONŒCIA.   *Order*—POLYANDRIA.

ALTHOUGH the Hazel never acquires the full
dimensions of a tree, it gives so decided a character
to most of our woods and hedges, that it requires
a specific notice among our most remarkable
forest-trees. It possesses too a peculiar claim on
our attention from being the only British tree
which in its wild state produces edible fruit.

The tree described by Pliny, under the name
of Avellana or Abellina, appears to have been
the variety familiar with us by the name of
Spanish-nut. It was introduced, he says, into
Greece from Pontus, whence it was called the
Pontic-nut, Avellana being a provincial term derived from the place where it was extensively
planted, now called Avellino, a city of Naples.
The wild European Hazel he does not mention,
although several modern authors quote from him
passages which refer not to this tree but to the
Walnut. The nuts sent by Jacob as a present
to his son Joseph in Egypt were in all probability
Pistachio-nuts, a kind of fruit which may justly
be reckoned among the finest productions of
Palestine, and therefore well worthy of being
associated with the other offerings. They are
about the size of the Hazel-nut, but of an oblong,
angular form ; the kernel is of a peculiar greenish

colour, and, though somewhat oily, has a very agreeable taste. All the Eastern versions of the Bible render the passage (Genesis xliii. 11) by Pistachio-nuts.

The Hazel was considered by the Romans as injurious to the Vine, and was not allowed to be planted in the vineyard. In the rustic festivals of the same people, the goat, which was also an enemy to the Vine by browzing on the young shoots, was roasted on a spit made of Hazel-wood. The Hazel grows wild in all the temperate climates of Europe and Asia, and is found in England at an elevation of 1600 feet.

Evelyn derives his family name from this tree; he says: " I do not confound the Filbert, Pontic, or Filbord, distinguished by its beard, with our foresters or bald Hasel-nuts, which doubtless we had from abroad, and bearing the names of Ave-lan, Avelin, as I find in some antient records and deeds in my custody, where my ancestors' names were written Avelan, alias Evelyn, generally." He also mentions several places which received their names from the abundance of these trees growing near them. " For the place," he says, " they all affect cold, barren, dry, and sandy grounds: mountains, and even rocky soils pro-duce them; they prosper where quarries of free-stone lie underneath, as at Hasellwry in Wilts, Haselingfield in Cambridgeshire, Haslemere in Surrey, and other places; but more plentifully, if the ground be somewhat moist, dankish, and mossy, as in the fresher hollows and sides of hills, hoults, and in hedge-rows." In the legendary history of the early English Church the Hazel stands beside the Whitethorn.*   " The most signal honour it

See Vol. I. p 184.

was ever employed in, and which might deservedly exalt this humble and common plant above all the trees of the wood, is that of hurdles, especially the flexible white, the red, the brittle; not for that it is generally used for the folding of our innocent sheep, an emblem of the Church, but for making the walls of one of the first Christian oratories in the world; and particularly in this island, that venerable and sacred fabrick at Glastonbury, founded by St. Joseph of Arimathea, which is storied to have been first composed but of a few small Hasel-rods interwoven about certain stakes driven into the ground; and walls of this kind, instead of laths and punchions, superintended with a coarse mortar made of loam and straw, do to this day inclose divers humble cottages, sheds, and out-houses in the country."

The Hazel was formerly, and indeed in some of the mining districts of England is still, believed to have an affinity for metals, being employed in the discovery of mines. The professor of this questionable science, as it was deemed, selected for this purpose a forked Hazel-rod (called a dowsing rod), a branch of which he held with each hand in front of his chest, with the other end slightly pointing outwards. He then walked forward over the ground to be examined, and when he reached a spot, under which there lay a lode or mass of metal, the end of the rod, in spite of his utmost efforts to restrain it, bent down, and pointed towards the buried mineral. The art, or rather imposture, is practised with very little alteration at the present time, in some parts of Cornwall, and mines are sunk in very unpromising places, solely on the strength of predictions derived from the divining rod. The virtue of this mysterious instrument is not confined to the pointing out of metal in its natural state, but extends to the discovery of hidden treasures. The effect is not supposed to depend on any exercise of witchcraft, but on a natural sympathy between the rod and metallic substances; and strange to say, both the practitioner and his deluded employers are frequently men who have correct scientific views in other respects. Still more wonderful properties were attributed to the Hazel in Evelyn's time; but he expresses himself on the subject very cautiously:— "Lastly, for riding switches, and divinatory rods for the detecting and finding out of minerals; at least, if that tradition be no imposture. By whatsoever occult virtue the forked stick, so cut and skilfully held, becomes impregnated with those invisible steams and exhalations, as by its spontaneous bending from an horizontal posture, to discover not only mines and subterraneous treasure, and springs of water, but criminals guilty of murder, &c., made out so solemnly, by the attestation of magistrates, and divers other learned and credible persons, who have critically examined matters of fact, is certainly next to a miracle, and requires a strong faith."

The usual form of the Hazel in its wild state, is a straggling bush, consisting of a number of long flexible stems from the same root. The bark on the young branches is ash-coloured and hairy, that on the old stems mottled with bright brown and grey. The leaves are rounded, stalked, and rough, and furnished at the base with oblong stipules, which soon fall off. The flowers are among the very earliest harbingers of returning

spring, reminding us that though winter, is the
season of rest with the vegetable world, that rest
is not the sleep of death. Almost before the
Snowdrop has ventured to peep out from its icy
home, the nut trees are plentifully decorated with
their yellow catkins, and if we search very closely,

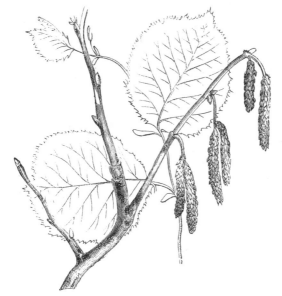

FLOWERS AND FOLIAGE OF HAZEL.

we shall find, towards the end of January, the
crimson pistils of the fertile flowers timidly push-
ing forth from some of the scaly buds, not less
beautiful than the more conspicuous catkins,
though scarcely known to any but the all observing botanist. The former, as soon as they have

shed their pollen, turn brown and fall off;
the latter, too, disappear, but in the course of a
few months may be detected, as bunches of nuts,
hiding themselves under the now fully expanded
foliage. At this season, a beautiful little beetle,
 balaninus nucum, guided by a
mysterious instinct, pierces the
yet tender shell of the nut,
and lays a single egg. The
soft pithy substance which it
contains, not being adapted for the sustinence of
the grub, the egg remains without undergoing any
change for some weeks ; but when the kernel has
nearly acquired its full size, a small white grub is
hatched, which immediately begins to feed on the
nut, and when full-grown, shews that although its
sole food has hitherto been of the softest kind,
it is provided with a powerful apparatus for gnaw-
ing a very hard substance. About the time that
the nut is ripe, the insect prepares for a change
of habitation, by boring a hole through the shell
and forcing its way out. It then falls to the
ground and buries itself in the earth, where it
constructs a cell, and is changed into a pupa, and
in the following season comes forth as a perfect
insect. We may well wonder at the instinct
which directs this little beetle to choose from
among all the trees of the forest, the one which
alone will afterwards bear abundance of food for
its offspring, and food too which it never eats
itself; and it is no less remarkable that it appears to know if the nut has been already occupied by some other insect of the same kind, for
we never find two grubs inclosed in the same shell.
It can have gained its knowledge neither by expe-

rience nor by education; for it lives but a single year in its perfect state, and it can have had no communication with others wiser than itself, for all are equally ignorant of their own history. We can therefore only conclude that in all its operations it has been guided by an intelligence superior to its own, by Him, namely, whose care is equally bestowed on the minutest and on the most important of His works.

The larvæ of other insects feed on the nut; but the depredations committed by squirrels, where these beautiful but mischievous little animals abound, exceed those of all the others. The food of the squirrel varies with the seasons; in winter and spring it feeds on buds and the bark of trees, and is said also to devour insects. In plantations of Larch, it often does great mischief by stripping off the bark from the young branches, and checks the growth of the trees by destroying the leading shoots. As soon as the kernel of the nut begins to swell, it makes this its principal food, and from July to October enjoys many a dainty repast. So eager is it in its search after nuts, that it will resort to trees growing close to dwelling-houses, and unless scared away (which is no easy task) will appropriate a large proportion of the fruit to its own use. If it only attacked the ripe nuts, its ravages would be limited, and perhaps be compensated, by the activity and intelligence displayed in its movements; but as it sets to work from six to eight weeks before the nuts are ripe, and destroys more than it actually devours, its share in the produce is more than an equitable one. The annexed woodcut represents a bunch of nuts which has been visited by a

squirrel. The depredator does not waste his
strength by cutting through the stem, but having
first nibbled away the husk, gnaws a hole through
the shell, and extracts the kernel piece-meal. If
the nut should happen to fall off before it is consumed, he does not take the trouble to descend in
quest of it, but begins upon another, and proceeds
until his voracious appetite is satisfied. Not un-
frequently a nut falls in his way, the kernel of

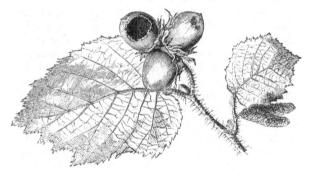

HAZEL-NUT.

which is not matured; this he either avoids altogether, or commences nibbling, but finding, probably from the hollow sound emitted, that his
labours will not be rewarded, he deserts it before
he has pierced through the shell. This power of
detecting a worthless nut appears to be gained by
experience; for we sometimes, though very rarely,
find a hollow nut, the shell of which has been
perforated.

The Nuthatch displays no less ingenuity than
the squirrel in procuring a meal from the Hazel-

tree. It has a strong and powerful beak, but having no means of holding its food, like the squirrel, while at work on the shell, it gathers the nut by the stem, and carries it away in its mouth to some rough barked tree, generally an Oak, strips off the husk, and fixes the nut in

THE NUTHATCH.

an angular crack in the bark, always selecting, as far as I have observed, a fissure so shaped that every blow which it deals with its beak wedges the nut more firmly; it thus cracks the shell and regales itself on its contents. In the months of July and August, when the woods are quieter

than at any other season of the year, a succession
of loud and quickly repeated tappings is a certain

NUT IN BARK.

guide to its haunts. I am
even inclined to think that
the noise which it makes is
understood by the squir-
rel, for the latter animal
frequently resorts to gardens in quest of filberts at
this season, though at other
times it is very shy, and
confines itself to woods
and plantations. In the
Midland Counties, an Oak
standing in a Hazel-copse
generally has the remains
of a few nuts wedged into its bark, and later
in the season, acorns may be found similarly
placed.

But in spite of squirrels and nuthatches, every
Hazel-copse will retain enough clusters to tempt
young people to " go a-nutting," a pastime which
is perhaps more delightful than any other of the
many which the country offers. The 14th of September appears to have formerly been the day
generally selected for this amusement. Brand
cites from the old play of " Grim, the Collier of
Croydon:"—

> " This day, they say, is called Holy-rood day,
>   And all the youth are now a nutting gone."

It appears, too, from a curious old manuscript
relating to Eton-school, that in the month of
September, " on a certain day," most probably
the fourteenth, the scholars there were to have a

play-day, in order to go out and gather nuts, a
portion of which, when they returned, they were
to make presents of to the different masters; but
before leave was granted for their excursion, they
were required to write verses on the fruitfulness
of autumn and the deadly cold of the coming
winter.

Thomson's allusion to this sport is well known:

> " Ye swains, now hasten to the Hazel-bank,
>   Where down yon dale, the wildly winding brook
>   Falls hoarse from steep to steep. In close array,
>   Fit for the thickets and the tangling shrub,
>   Ye virgins come. For you their latest song
>   The woodlands raise ; the clustering nuts for you
>   The lover finds amid the secret shade ;
>   And where they burnish on the topmost bough,
>   With active vigour crushes down the tree,
>   Or shakes them ripe from the resigning husk,—
>   A glossy shower, and of an ardent brown,
>   As are the ringlets of Melinda's hair."

Wordsworth's lines on the same subject are so
beautiful, and recall so forcibly feelings which
every one must have experienced who has spent
his early years in the country, that I cannot for-
bear quoting them at length:—

> " ——— It seems a day
>   (I speak of one from many singled out)
>   One of those heavenly days that cannot die ;
>   Where in the eagerness of boyish hope,
>   I left our cottage threshold, sallying forth
>   With a huge wallet o'er my shoulders slung,
>   A nutting crook in hand ; and turned my steps
>   Towards some far-distant wood, a figure quaint,
>   Tricked out in proud disguise of cast off weeds,
>   Which for that service had been husbanded,
>   By exhortation of my frugal dame—
>   Motley accoutrement, of power to smile
>   At thorns, and brakes, and brambles,—and, in truth,
>   More ragged than need was ! O'er pathless rocks,

Through beds of matted fern and tangled thickets,
Forcing my way, I came to one dear nook
Unvisited, where not a broken bough
Drooped with its withered leaves, ungracious sign
Of devastation ; but the Hazels rose
Tall and erect, with milk-white clusters hung,
A virgin scene !   A little while I stood,
Breathing with such suppression of the heart,
As joy delights in ; and, with wise restraint
Voluptuous, fearless of a rival, eyed
The banquet ;—or beneath the trees I sate
Among the flowers, and with the flowers I played :
A temper known to those, who, after long
And weary expectation, have been blest
With sudden happiness beyond all hope.
Perhaps it was a bower, beneath whose leaves
The violets of five seasons re-appear,
And fade unseen by any human eye ;
Where fairy water-brakes do murmur on
For ever ; and I saw the sparkling foam,
And, with my cheek on one of those green stones
That, fleeced with moss, beneath the shady trees,
Lay round me, scattered like a flock of sheep,
I heard the murmur and the murmuring sound,
In that sweet mood when pleasure loves to pay
Tribute to ease ; and, of its joy secure,
The heart luxuriates with indifferent things,
Wasting its kindliness on stocks and stones,
And on the vacant air.   Then up I rose,
And dragg'd to earth both branch and bough, with crash
And merciless ravage : and the shady nook
Of Hazels, and the green and mossy bower,
Deform'd and sullied, patiently gave up
Their quiet being : and unless I now
Confound my present feelings with my past,
Ere from the mutilated bower I turned
Exulting rich beyond the wealth of kings,
I felt a sense of pain when I beheld
The silent trees, and saw the intruding sky."

On the banks of the river Plym, in Bickleigh
Vale, Devonshire, stands the largest and most
productive Hazel that I have ever seen.   Owing,
however, either to its extreme age, or more probably to its roots extending to a depth below the

level of the river, its promising clusters of nuts
never contain kernels, but are filled with a black
spongy substance, and are consequently worthless.
The tree stands alone, and overspreads a considerable extent of green sward, which in the nutting-
season is pretty sure to be thickly strewed with
clusters of large and very fresh-looking nuts. A
favourite practical joke with a person acquainted
with this deceitful tree is to lead a nutting-party
hither, as if carelessly, and while all are congra-
tulating themselves that they will now be able to
fill their baskets without further trouble, to stand
aloof until the ground is clear, and then to recommend his companions to test the value of their
easily acquired store, by tasting a few. Many a
long face have I seen under this tree, as the disappointed collectors have carefully picked out
from their wallets the now despised clusters, and
scattered them, in the hope of practising a similar
imposition on the next foraging party who might
happen to come up. Trees standing near either
running or stagnant water should be looked on
suspiciously, for though the Hazel attains a large
size in such situations, its produce is generally
worthless.

Phillips mentions his having gathered nuts "from
a Hazel which grew on the top of the church-
steeple at Henfield in Sussex. Mr. Borrer informs me that the parish-clerk of Henfield well
remembers the tree in question, and says that he
himself (a bricklayer by trade) assisted in destroying it when the church-tower was under
repair nearly fifty years ago. He describes it
as of considerable size, growing at the top of one
of the buttresses, and producing plenty of nuts,

II.   L

which the people used to pick up at the bottom
of the tower, its situation not admitting of their
being gathered either from the bottom or from
the top. On one occasion two boys, brothers,
determined, at all hazards, to get at the nuts; they
accordingly mounted to the top of the tower, one
of them held the other by the feet over the battlements, and the latter gathered them, though it
can scarcely be said at his ease. If both these
stories are correct, Phillips himself must have
been the venturesome youth, and perhaps was
ashamed, when he had grown older and wiser, to
mention that he had ventured his neck for so mean
a booty as a handful of nuts.

Nuts were, in ancient times, in great demand
on Allhallow Eve, Oct. 31st, which, from that
circumstance, was sometimes called " Nut-crack
Night." A nut was chosen to bear the name of
each unmarried person in the company, and placed
close to the fire until it ignited; and it was pre-
tended that the way in which it burned prognos-
ticated certain events in the life of the person
whose name it bore. Burns says that the same
custom was observed in Scotland; and in Ireland,
this and other antiquated customs sometimes afford
amusement to parties of young people at the present day.

The Hazel rarely attains such a size as makes
it important in the landscape; it is nevertheless
valuable when fulness of foliage is desired, retaining its leaves until almost every other tree has
been dismantled, and assuming a bright warm
yellow before they fall, which gives to autumn a
lingering beauty, that it would otherwise want.
Even when the leaves have fallen, the tree is not

bare; for the barren catkins expand almost immediately after, and remain in flower all the winter.

Dr. Plot relates, in his " Natural History of
Oxfordshire," that some workmen digging a pit at
Watlington Park, found, at a depth of fifty or
sixty feet, a large number of entire Oak-trees,
lying in confusion, and "all along as they dug,
they met with plenty of Hazel-nuts, from within
a yard of the surface to the bottom of the pit,
which time's iron teeth had not yet cracked; and
that which amazed me most of all, I think they
lay thicker than ever they grew. The shells of
the nuts were very firm without, but nothing remained within of a kernel, but a show of the dry
outer rind."

A still more remarkable discovery of nuts was
made about twenty or thirty years since at Car-
rickfergus, County Antrim, Ireland. These were
found in great numbers, and at various depths on
the sea-shore: the husk, in all that I examined,
had disappeared; the shell was much softer than
in recent specimens, and liable to crack, unless
kept in water, and the kernel was converted into
a whitish semi-opaque stone. They were decidedly of the same species as the common Hazel-
nut, and indeed were only to be distinguished
from the old nuts, which one commonly finds on
the ground in Hazel-copses, by their superior
weight. How they came into this situation and
were subsequently submitted to a partial conver-
sion into stone, are questions which have not
satisfactorily been accounted for.

The Hazel is rarely found of a sufficient size
to supply building materials; but the young rods
being tough and flexible, are much used for

hoops, walking-sticks, fishing-rods, &c.; and from their smoothness and pleasing colour they are well adapted for making rustic seats, and tables for summer-houses. For this purpose they are split, cut to a suitable size, and nailed, in various patterns, to smooth boards of some other wood. They are also excellent as fire-wood, and when converted into charcoal make the best gunpowder. The charcoal crayons, used by artists for drawing outlines, are also prepared from Hazel-wood."

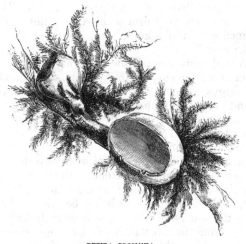

PEZIZA COCCINEA.

One of the most beautiful of the British fungi (*Peziza coccinea*) grows on decaying branches of the Hazel, and may be found lying on the ground in damp places from December to April. In their early stage they are whitish, club-shaped columns:

but soon the summit opens and exposes an intensely bright crimson surface, which expands at first into a deep cup, and finally into a spreading bowl, as large over as a crown-piece. When in this state, if they are touched while the sun is shining warmly, they will sometimes send up a fine jet of smoke, at least so it is in appearance, but so rapid is this process, that before one has had time to discover from what part of the surface the puff proceeded, it has vanished, and not a pore, as large even as the point of a needle, can be detected. The particles of which this apparent smoke is composed are, undoubtedly, seeds; but how infinitely minute, and yet how incalculably numerous must they be, that they should vanish from the sight too rapidly for the eye to follow them, and yet exist in such numbers as to be visible at all!

The principal varieties of Hazel cultivated in Great Britain are the Filbert and Cob-nut; the former of which is distinguished by its ample husk, which entirely conceals the nut, the latter by its larger size. In mode of growth and character of foliage, neither of these differs materially from the Common Hazel. The name Filbert, is supposed to be a corruption of *full beard,* from the lengthened appendage to the nut, an etymology, which, unsatisfactory as it is, is the only plausible one which has been given. Considerable skill is requisite in the cultivation of the Filbert, in order to ensure an abundant crop. In the neighbourhood of Maidstone, in Kent, where they are grown in greater abundance and perfection than anywhere else in England, the trees are trained with short stems, like gooseberry-bushes, and are

pruned into the shape of a bowl, very thin of
wood, and never exceeding six feet in height.

THE FILBERT.

The produce from trees thus treated is in certain
situations enormous; as much as a ton and a half
having been gathered from a single acre: a ton
an acre is, however, considered a large crop, and
as total failures are very common, five hundred
weight per acre is considered a fair average.   The

treatment of the Cob-nut is the same as that of
the Filbert-tree.

THE COB-NUT.

The nuts exposed for sale, under the name of
Barcelona or Spanish nuts, are the produce of a
tree differing little from the varieties known in
this country. They are imported from different
parts of France, Portugal, and Spain, and especially from Tarragona, in the last-named coun-

try; from whence no less than 25,000*l*. worth are annually exported for the English market alone. M'Culloch states, that the entries of nuts (from all quarters) for home consumption, amount to from 100,000 to 125,000 bushels a-year.

THE WALNUT.

THE WALNUT.

JUGLANS REGIA.

Natural Order—JUGLANDACEÆ.

Class—MONŒCIA.   *Order*—POLYANDRIA.

THIS noble tree, though not a native of Europe, was extensively cultivated in Greece and Italy, at a very early period. Its most ancient names were Persicon (Persian-tree), and Basilicon (King-ly-tree), both indicating its eastern origin. The Greeks also called it Caryon, from *kara* a head, because its powerful odour was supposed to cause headache, or from some fancied resemblance between the nut and the human head. The Romans, to mark the estimation in which they held it, gave it the name of Juglans, or Jupiter's mast, from its being as much superior to other kinds of mast, as their false god was supposed to be superior to men.

Its shade was thought, in Pliny's time, to be injurious not only to the human body but to all kinds of vegetables; nevertheless its nuts were highly prized both as an article of food, and for numerous medicinal properties, especially as an antidote to poison and the bite of a mad dog. The husk of the nut was used as a dye, and an oil was expressed from the kernel, which was also considered a valuable medicine.

It was customary at weddings in Rome for the bridegroom to throw about handfuls of nuts to be scrambled for by boys, as a sign that he had now laid aside childish amusements, a custom to which frequent allusion is made by the Latin poets.

The Walnut is a large spreading tree, with a rough trunk, and strong crooked branches, which diverge from the main stem, somewhat after the manner of the Oak. The leaves are pinnate, like those of the Ash, but much larger : when young they are tinged with red, and at all periods until

FLOWERS AND FOLIAGE OF WALNUT.

they wither, emit a powerful and fragrant perfume when slightly bruised. The young branches are brittle, and remarkably stout to the very extremity; the bark on these is smooth and shining. The rudiments of the barren flowers appear early in the summer previous to their expansion, and are conspicuous in the axils of the leaves, in the shape of short conical spurs, which are smooth, and of a greenish brown hue. In the following summer these lengthen into drooping, cylindrical catkins. The fertile flowers do not shew themselves before the year that they bear fruit, when they appear among the leaves at the extremities of the shoots, and are at no time so conspicuous as the barren flowers.

TWIG OF WALNUT.

The poet Virgil remarks, that when the Walnut produces an abundance of blossom, a good corn-harvest may be expected, and that the reverse will be the case when it bears a profusion of leaves and few flowers. Agricultural maxims of this kind are frequently founded in truth, but I am

not aware whether or not there are any grounds
for considering this opinion correct. The fertile
flowers are succeeded by bunches of smooth nuts,
which in their young state are firm but not hard,
and abound in juice of a strong and offensive
odour, which, on exposure to the air, turns dark
yellow, and subsequently black. In September
or October the outer case becomes mealy, and
splits irregularly, disclosing the nut, which is too
well known to need any description. The latter
then falls off, generally bringing the case with it.
About the same time the leaves turn yellow and
fall off, and the tree is more than ever marked by
its wide-spreading crooked arms, and its thick
terminal branches, plentifully furnished with the
conical flower-buds described above.

The Walnut, besides being a native of Persia,
grows wild in Tartary, where a single tree is
said to produce as many as from forty to sixty
thousand nuts yearly; and we learn from Dr.
Clarke that the Tartars pierce the Walnut-trees
in the spring, when the sap is rising, and put in a
spigot for some time; and that when it is with-
drawn, a clear sweet liquor flows out, which
when coagulated, they use as sugar. The tree was
introduced into Europe at an early period, and
probably passed into Britain from France, the
first syllable of the word Walnut being a corruption of "Gaul," in accordance with an ety-
mological change observable also in the words
" Wales" and " Cornwall."

" It delights," says .Evelyn, " in a dry, sound,
and rich land, especially if it incline to a feeding
chalk or marl; and where it may be protected
from the cold (though it affects cold rather than

extreme heat,) as in great pits, valleys, and high-way-sides; also in stony grounds, if loamy, and on hills, especially chalky; likewise in corn-fields. Thus Burgundy abounds with them, where they stand in the midst of goodly wheat-lands, at sixty and an hundred feet distance; and so far are they from hurting the crop, that they are looked upon as great preservers, by keeping the ground warm; nor do the roots hinder the plow. Whenever they fell a tree, which is only the old and decayed, they always plant a young one near him; and in several places betwixt Hanaw and Frankfort, in Germany, no young farmer what-soever is permitted to marry a wife, till he brings proof that he hath planted, and is a father of such a stated number of Walnut-trees, and the law is inviolably observed to this day, for the extra-ordinary benefit which this tree affords the in-habitants.

"They render most graceful avenues to our country dwellings, and do excellently near hedge-rows; but had need to be planted at forty or fifty feet interval, for they affect to spread both their roots and branches. The Bergstras (which ex-tends from Heidelberg to Darmstadt,) is all planted with Walnuts; for so, by another ancient law, the Bordurers were obliged to nurse up and take care of them, and that chiefly for their ornament and shade, so as a man may ride for many miles about that country under a continued arbor or close walk, the traveller both refreshed with the fruit and the shade."

In Evelyn's time there were extensive plan-tations of Walnuts, particularly on the downs near Letherhead in Surrey, at Godstone, and at

Carshalton, "where many thousands of these do celebrate the industry of the owners;" and this is still the case in many parts of the continent. In the south of France especially, the fruit, oil, and wood form some of the principal articles of commerce; and here, as well as in the north of Italy and in Switzerland, the roads are lined for miles together with Walnut-trees. During August and September, when the fruit is ripe or nearly so, and the weather so warm that the shelter of a house is not required to protect the traveller from cold, he may walk under the shade of the tree, eating the fruit by day and sleeping under it by night.

Walnuts in their young state are both pickled and preserved. For this purpose they should be gathered at the end of June or beginning of July. If intended for pickling, they should be soaked in salt and water for a fortnight before they are placed in the vinegar. "They may be preserved," says Loudon, "either with or without their husks; in the latter state they are the most agreeable, but in the former most strengthening to the stomach." Gerard says, "the green and tender nuts, boyled in sugar, and eaten as suckarde (sweetmeat) are a most pleasant and delectable meate, comfort the stomache and expell poyson." A fine stomachic liqueur is made from the young nuts about the beginning of June; and in August, before the shells become hard, they are eaten, what the French call, *en cerneaux*, that is, with the kernel while green scooped out with a short brass knife, and seasoned with vinegar, salt, pepper, and shallots. When ripe, they are considered wholesome as long as the skin can easily be separated from the kernel, soon after which they become oily and indigestible. When they have been kept for a few months, they are in a fit state to be converted into oil, which is either used for culinary purposes and burning, or, more generally, is employed by artists in the preparation of fine colours; it is preferred to any other kind of oil for this purpose, on account of its fluidity and the rapidity with which it dries.

As a timber-tree, the Walnut holds a high rank: in young trees the wood is white and comparatively soft; but in full-grown trees it becomes compact, and of a dark brown colour, beautifully veined and shaded with light brown and black. Before the discovery of mahogany it was much used for furniture, and many a curiously wrought cabinet or book-case is still to be found in old-fashioned houses; its principal use, however, at the present time is for gunstocks, for which it is admirably adapted, combining the necessary qualities of lightness and strength, and being at the same time not liable to warp. "It is a remarkable fact in the history of this tree," says Loudon, "that in the winter of 1709, the greater part of the Walnut-trees of Europe were killed, or so far injured as to render it advisable to fell the trees. The Dutch at that time, foreseeing the scarcity of Walnut timber that was likely to ensue, bought up all the trees that they could procure in every direction, and sold them again according to the demand, for many years afterwards at a greatly advanced price.

During the wars of Napoleon Buonaparte, the demand for Walnut-timber became so great and the price rose so high, that 600*l.*

are said to have been given in England for a single tree.

The juice of the Walnut-tree, both that derived from the leaves and the husk, especially the latter, imparts a rich brown stain. Gypsies use this to dye their skin, and it is also employed in the staining of floors, to which it is desired to give a dark hue and a high polish.

The Walnut-tree sometimes produces a considerable quantity of manna; and it has been observed in France, that whenever the trees happen to yield more than ordinary, they usually perish the following winter.

The largest Walnut-trees in England are from fifty to sixty feet high, with a diameter of trunk from four to five feet, and that of the branches from sixty to ninety. The only tree to which any legendary interest is attached, is mentioned in Collinson's "Somersetshire." " Besides the Holy Thorn, there grew in the Abbey churchyard of Glastonbury, on the north side of St. Joseph's chapel, a miraculous Walnut-tree, which never budded forth before the feast of St. Barnabas, namely, the eleventh of June, and on that very day shot forth leaves, and flourished like its usual species. This tree is gone, and in the place thereof stands a very fine Walnut-tree of the common sort. It is strange to say how much this tree was sought after by the credulous; and, though not an uncommon Walnut, Queen Anne, King James, and many of the nobility of the realm, even when the times of monkish superstition had ceased, gave large sums of money for cuttings from the original."

THE LIME-TREE.

THE LIME-TREE.

TILIA EUROPÆA.

Natural Order—TILIACEÆ.

Class—POLYANDRIA.   *Order*—POLYGYNIA.

THE Lime or Linden-tree was well known to the Greeks under the name of Philyra; and the Romans, Pliny tells us, held it in great repute for its " thousand uses." The timber was employed in making agricultural implements, and was also considered to be well adapted for shields, as it was said to deaden the blow of a weapon better than any other kind of wood. Pliny states also, that it was not liable to be worm-eaten. The bark was a common writing material, and when split into ribands was made into head-dresses, which were worn on festive occasions. In medicine its supposed virtues were very great; the leaves and bark had a healing power, and decoctions of various parts beautified the skin and promoted the growth of the hair. The seed was said to be eaten by no animal. Evelyn mentions that a book written on the inner bark of the Lime, " was brought to the Count of St. Amant, governor of Arras, 1662, for which there were given eight thousand ducats by the Emperor; it contained a work of Cicero, *De ordinandâ Republicâ, et de inveniendis orationum exordiis;* a piece inestimable, but never pub-

lished, and now in the library at Vienna, after it
had formerly been the greatest rarity in that of
the late Cardinal Mazarin."

In the Middle Ages the same honours were paid
to the Lime-tree which belonged to the Poplar,
a tree which derived its name from being the
emblem of popular freedom. During the strug-
gles of the Swiss and Flemish to recover their
liberty, it was their custom to plant a Lime-
tree on the field of every battle that they gained
over their oppressors; and some of these trees,
particularly those planted by the Swiss in com-
memoration of their victories over Charles the
Bold, are still remaining, and have been the
subject of many ballads. " At Fribourg," Loudon
informs us, " there is a large Lime, the branches
of which are supported by props of wood. This
tree was planted on the day when the victory of
the Swiss over the Duke of Burgundy, Charles
the Bold, was proclaimed, in the year 1476; and
it is a monument admirably accordant with the
then feebleness of the Swiss Republics, and the
extreme simplicity of their manners. In 1831,
the trunk of this tree measured thirteen feet,
nine inches, in circumference." Another tree
stands near the same place, which is supposed to
be nearly a thousand years old; its trunk is
thirty-six feet in circumference and is still perfectly sound.

When too we recollect that the father of
modern botany, Linnæus, derived his name from
the Swedish Lin (our Linden-tree) we must allow
that it is recommended to us by the most pleasing
associations.

The Lime-tree occurs in Europe, under three

forms,* which are distinguished principally by the
size and smoothness (or the reverse) of their leaves.
They are all natives of the middle and north of
Europe, but the small-leaved species alone is
considered to be indigenous to Britain. Though
all these kinds have long become naturalized, we
rarely see them growing in places where there is
no room for suspicion that they may have been
originally planted; yet there is, in the neighbourhood of Worcester, on the authority of Mr.
Edwin Lees, a wood, remote from any old dwelling or public road, of above five hundred acres
in extent, the greater part of the undergrowth of
which is composed of the small-leaved Lime.
There are also in the same part of the country,
trees estimated to be upward of three hundred
years old.

The Lime is a large tree, characterized by its
pyramidal shape, by the multiplicity of its long,
slender, and upright branches, which start from
the main stem not many feet from the base, and
by the unbroken surface presented by its abundant foliage. These characters give to half-
grown trees, in which they are most conspicuous,
a stiff and formal appearance, especially if they
happen to be planted in rows. In older specimens, the weight of the lower branches frequently bends them down to the ground, so as
entirely to conceal the trunk; the middle part of
the tree is thus thrown open, and the pyramidal
outline destroyed; the summit too becomes
somewhat more tufted. Under these circumstances the Lime is a stately and even picturesque tree, especially when standing alone

* *Tilia Europæa, T. platyphylla,* and *T. parvifolia.*

or in groups of three or four on a sloping lawn. It is very patient of clipping, and, consequently, in the suburbs of large towns it more frequently disfigures than adorns, sometimes appearing as a mere leafy hedge, unmeaningly elevated on equi-distant columns.

The leaf is bright green, pointed, and heart-shaped at the base, smooth above, and either uniformly downy beneath, or bearing small tufts of down in the angles of the veins.

LEAF AND FLOWER OF THE LIME-TREE.

The flowers are scarcely less profuse than the leaves, and rendered very conspicuous by large yellowish-green bracteas, from the centre of which spring three or more stalked flowers. These consist of a five-parted calyx, and five petals, which are nearly of the same colour as the brac-

teas. The stamens are numerous, and the whole
flower is deliciously fragrant, especially towards
evening,
—" At dewy eve
Diffusing odours."

The seed-vessels are globular and downy, but
rarely perfect the seeds in England. While the
Lime-tree is in flower, it is frequented by myriads of bees, which

" Sit on the bloom, extracting liquid sweets
Deliciously."

Honey from the Lime is considered superior to
all other kinds for its delicacy of flavour; it is to
be obtained in a perfectly pure state only at the
little town of Kowno in Lithuania, which is surrounded by an extensive forest of Lime-trees.
The pleasing sound produced by the busy collectors, joined to the fragrant perfume diffused
by the flowers, frequently gives occasion for its
being planted near houses, in preference to other
more picturesque trees. Even after the flowers
have faded and fallen to the ground, the odour is
perceptible, the ground remaining for a long while
thickly strewed with the withered stamens, which
retain their fragrance to the last. Towards the
end of September the leaves turn to a bright
yellow, and in the course of the following month
fall off.

The custom of making avenues of Lime-trees
was adopted in the time of Louis XIV., and
accordingly the approaches to the residences of
the French as well as the English gentry of that
date, were bordered with Lime-trees. It subsequently fell into disrepute for this purpose, on

account of its coming late into leaf, and shedding
its foliage early in autumn, and was supplanted
by the Hornbeam and Elm; but many of the
cities of continental Europe still boast of their
public walks of Lime-trees, which in the hours
of relaxation are numerously frequented by
persons of all classes and ages. The Dutch,
especially, plant them in lines along their widest
streets, and by the sides of their canals, and the
whole country is perfumed by them during the
months of July and August. Evelyn, in whose
time straight walks and formally grown trees
were in vogue, recommends the Lime as "of all
other, the most proper and beautiful for walks,
as producing an upright body, smooth and even
bark, ample leaf, sweet blossom, the delight of
bees, and a goodly shade at the distance of eighteen or twenty-five feet."

The Lime-tree, though not applied to so many
uses as it was in the time of Pliny, is valuable
for many purposes. In the Belgian Horticultu-
rist it is stated, that "the flowers infused in cold
water are antispasmodic; and in hot water they
make an agreeable kind of tea. The leaves and
young shoots are mucilaginous, and may be employed in poultices and fomentations. The timber is better adapted than any other for the
purposes of the carver; it will take any form
whatever; it admits of the greatest sharpness in
the minute details, and it is cut with the greatest
ease. It is also used for sounding boards for
pianos and other musical instruments. But the
peculiar use of the Lime is for the formation of
mats from its inner bark. In June, when the
leaves begin to develope themselves, and the tree

is full of sap, branches or stems of from eight
to twenty years' growth, are cut and trimmed, and
the bark is separated from them from one end to
the other. This is easily done, by simply drawing the edge of a knife along the whole length of
the tree or branch, so as to cut the bark to the
soft wood. It then rises on each side of the
wound, and almost separates of itself. If mats
are to be made immediately, the bark is next
beaten with mallets on a block of wood, and children are employed to separate the inner bark,
which comes off in strands or ribands, while the
outer bark detaches itself in scales. If mats are
not to be made for some time, the bark is dried
in a barn or shed, and either kept there, or stacked
till it is wanted. It is then steeped twenty-four
hours in water, beaten as before, and put into a
heap, where it remains, till it undergoes a slight
fermentation. When this has taken place, the inner
bark separates in ribands and shreds as before.
With the shreds, cords of different kinds are
twisted in the usual manner; and mats are formed
with the ribands in the same way as rush mats.
The ribands which are to be used in forming
mats for gardens undergo a sort of bleaching for
the purpose of depriving them of part of their
mucilage, which would otherwise render them
too liable to increase and diminish in bulk by
atmospheric changes. The great advantage of
Lime, or bast mats, over all others in gardens, is
that they do not so easily rot from being exposed
to moisture."

The superiority of Lime wood for the purposes
of sculpture, is confirmed by the fact that Gibbon, the celebrated carver in wood, preferred it

to any other. This remarkable person was first introduced to public notice by Evelyn, the author of the " Sylva," himself a man who, whether as a churchman, a citizen, or a man of taste, may serve as a model to his countrymen. The following extract from his Diary will be read with interest :—

" 1670-1. 18th Jan.—This day I first acquainted his Majesty with that incomparable young man Gibbon, whom I had lately met with in an obscure place, by mere accident, as I was walking near a poor solitary thatched house, in a field in our parish, near Say's Court. I found him shut in; but looking in at the window I perceived him carving that large cartoon or crucifix of Tintoret, a copy of which I had myself brought from Venice, where the original painting remains. I asked if I might enter; he opened the door civilly to me, and I saw him about such a work as for the curiosity of handling, drawing, and studious exactness, I had never before seen in all my travels. I questioned him why he worked in such an obscure and lonesome place; he told me it was that he might apply himself to his profession without interruption, and wondered not a little how I had found him out. I asked if he was unwilling to be made known to some great man, for that I believed it might turn to his profit; he answered, he was yet but a beginner, but would not be sorry to sell off that piece; on demanding the price, he said 100*l.* In good earnest the very frame was worth the money, there being nothing in nature so tender and delicate as the flowers and festoons about it, and yet the work was very strong; in the piece was more than a hundred figures of men, &c. I found he was likewise musical, and very civil, sober, and discreet in his business. There was only an old woman in the house. So desiring leave to visit him some-times, I went away." The amiable patron was, however, disappointed in procuring a sale for the work: for when the sculptor was taken to Whitehall, " the King, being called away, left us with the Queen, believing she would have bought it, it being a crucifix, but when his Majesty was gone, a French peddling woman, one Madame de Boord, who used to bring petticoats and fans and baubles out of France to the ladies, began to find fault with several things in the work, which she understood no more than an ass or a monkey, so as in a kind of indignation, I caused the per-son who brought it to carry it back to the cham-ber, finding the Queen so much governed by an ignorant French woman, and this incomparable artist had his labour only for his pains, which not a little displeased me, and he was fain to send it down to his cottage again."

Walpole calls Gibbon " An original genius, a citizen of nature. There is no instance before him of a man who gave to wood the loose and airy lightness of flowers, and chained together the various productions of the elements, with the free disorder natural to each species." Many fine specimens of Gibbon's carvings still exist in their original beauty at Windsor Castle, St. Paul's Cathedral, and many of the mansions of the no-bility.

To the above-mentioned uses to which the Lime may be applied, Loudon adds the following. The Russian peasants weave the bark of the young shoots for the upper part of their shoes, the outer bark serves for the soles; and they also make of it baskets and boxes for domestic purposes. The fishermen of Sweden make nets for catching fish of the fibres of the inner bark, separated by maceration, so as to form a kind of flax; and the shepherds of Carniola weave a coarse cloth of it, which serves them for their ordinary clothing. The sap drawn off in the spring affords a considerable quantity of sugar, and the seed may be converted into an oily substance perfectly resembling chocolate, but unfortunately of little value, as it soon becomes rancid.

Among the many remarkable Lime-trees described by various authors, the following are most worthy of notice. At Chalouse, in Switzerland, there stood one in Evelyn's time, under which was "a bower composed of its branches, capable of containing three hundred persons sitting at ease; it had a fountain set about with many tables, formed only of the boughs, to which they ascend by steps, all kept so accurately, and so very thick, that the sun never looked into it." The same author mentions another famous Lime at Neustadt in Wirtemburg, which gave a distinctive name to the town. Its huge limbs were supported by numerous stone columns bearing inscriptions. This tree, Loudon tells us, was in 1838 still in existence, the trunk being eighteen feet in diameter, and the limbs being supported by 118 columns. The people of Neustadt are in the habit of sitting in this tree to eat fruit; and several gooseberry trees have sprung up in the crevices and hollows of the bark, the fruit of which is sold to visitors.

Phillips describes an enormous Lime standing in the village of Prelly, in the canton of Vaud in Switzerland, under the shade of which the village council formerly held their deliberations.

One of the finest Lime-trees in England stands in Moor Park, Hertfordshire, among many others of immense size. The circumference of the trunk is twenty-three feet and a half; at nine feet from the ground, it sends out nineteen horizontal branches (each of which would make a good sized tree), to a distance of from sixty to seventy feet. It is nearly a hundred feet high.

A yet more remarkable tree, described by Loudon, growing at Knowle, covers nearly a quarter of an acre of ground. The lower branches, which extend to a great length, have rested their extremities on the soil, rooted into it, and sent up a circle of young trees. The outer branches of this outer row of trees have in their turn rooted and thrown up a second row of trees, which were, in 1820, thirty feet high.

Several American species of Lime have been introduced into England; but none of these require a particular notice.

THE BARBERRY.

BERBERIS VULGARIS.

Natural Order—BERBERIDÆ.

Class—HEXANDRIA.　　*Order*—MONOGYNIA.

THIS pretty shrub grows wild in many parts of
England, and is of still more common occurrence
in gardens and shrubberies, where it is cultivated
for the sake of the pleasing appearance presented
by its numerous clusters of yellow flowers, and
drooping bunches of scarlet berries. It is indigenous to most of the countries of Europe and
Asia, and is also found in North America, preferring a temperate climate, but also inhabiting
warmer regions; in which latter case it grows at a
high elevation in the mountains.

In its wild state, in England, it appears in the
form of a low bushy shrub, but when cultivated,

BLOSSOM OF BARBERRY.

attains a height of twenty
feet or more. The branches
are covered with smooth bark
of a remarkably light hue,
and, with the three-forked
thorns, sufficiently distinguish
it from every other shrub, even
when it is bare of foliage. The
leaves are nearly elliptical, smooth, and beautifully
fringed at the edge. The flowers consist of a
calyx of six unequal leaves, and as many concave
yellow petals, in each of which is concealed a

stamen with a flattened filament. The flowers last only a short time, during which they are showy, but emit a very unpleasant odour. Near the base of each fila-ment is a small spot, which possesses a high degree of irritability. If this be touched by any small body, while the bloom is in perfec-tion, the stamen sud-denly bends forward and closes on the pis-til, and, if allowed to remain for a few hours, gradually returns to its original position, ready to perform the same movement, when again excited. It is a well-known fact that no flower will bear fertile seeds unless some portion of pollen be lodged on the pis-til, while the latter

FLOWERS AND FRUIT OF THE
BARBERRY.

is in its mature state. In the Barberry, when the flower is expanded, the anthers containing the pollen are bent so far away from the stigma or summit of the pistil, that they could scarcely perform their office, were they not by some means raised and brought forward, and that too in bright, sunny weather. This is just the time when insects are most busily occupied in exploring for food; and they, in their search after honey, visit the flowers of the Barberry, and cannot fail to touch some one or other of the stamens, which instantaneously springs forward from the shelter of its petal, and sheds a portion of the pollen on the pistil.   Thus, the apparent defects in the harmony of nature are, if looked into, found to be occasions of displaying the unerring wisdom of the Supreme Creator.

The berries of the Barberry are oblong, and when ripe, scarlet, and covered with a bloom like that of the plum.   They are intensely acid, so much so as to be refused by birds; they therefore remain a long time on the tree, and when produced in abundance are very ornamental.

The inner bark and wood are bitter and astringent, and of a bright yellow colour, which may be extracted, and furnishes good dye.   The leaves are acid, but are not now applied to any use. The berries, preserved in various ways, are made into jelly, comfits, cooling drinks, and pickles. For these purposes a variety is preferred which bears seedless berries.

A notion was formerly prevalent that the Barberry caused mildew or rust in corn, and consequently many persons destroyed it whenever it was found growing near arable land.   Botanists, however, have sufficiently proved that the orange-coloured mildew, which infests the leaves of the Barberry, though nearly of the same colour as the mildew of corn, is totally different from it, and cannot be transferred to any other plant.*   It is,

---

* The mildew of wheat is not produced by a superficial fungus like an Erisyphe (the rust of the Barberry), but an intestinal fungus of the genus *Puccinia*, and consequently to place such leaves among wheat is not likely to injure it."—*Gardeners' Chronicle.*

therefore, to be hoped that the Barberry will be allowed to retain its place as a hedge-shrub, for which its habit of growth and numerous stout prickles admirably adapt it.

Several foreign species of Barberry are cultivated in gardens, some of which, from the north-west coast of North America, are among the most ornamental evergreen shrubs that have ever been introduced. These are placed by some botanists in a distinct genus, *Mahonia*, but with questionable propriety.

THE TAMARISK.

THE TAMARISK.

TAMARIX GALLICA.

Natural Order—TAMARISCINEÆ.

Class—PENTANDRIA.   *Order*—TRIGYNIA.

THE Tamarisk is a native of most of the countries of Southern Europe, Asia Minor, Tartary,
Japan, Barbary and Arabia, assuming a great
variety of forms, according to the soil, situation,
and climate in which it grows.   It was known to
the Greeks and Romans under the name of
Myrica, and frequent mention of it occurs in the
writings of the ancients.   Pliny describes it as
an evergreen, but this title it scarcely merits with
us, for it only partially retains its foliage during
the winter.   It is, however, a very pleasing shrub,
remarkable for the rich purple of its long tapering branches, and the light feathery appearance
of its spray.   The flowers are produced in July,
growing in bunches of spikes near the ends of the
shoots; they are flesh-coloured, with red stamens.

The Tamarisk is scarcely indigenous to Britain;
for though it is said by some to be wild in Cornwall and on other parts of the coast, it bears every
appearance of having been planted.   It was first
observed in an apparently wild state on St.
Michael's Mount, whither, perhaps, it may have
been brought from the opposite coast of France
by smugglers.   It is now a common hedge-plant
in many parts of the Cornish coast, having been
introduced, it is said, into the Lizard district by

a carter, who having lost his whip, gathered one
of the long flexible branches at the Mount, and
at the conclusion of his journey stuck the rod

FLOWERS OF THE TAMARISK.

into the ground, where it grew, and was soon
extensively propagated. It is far from improba-

ble that it was introduced in some such way from
France, for it grows from cuttings as freely as
the Willow, provided that it be planted in
autumn or early in spring. On the continent, it
is said to grow in the greatest abundance on the
banks of rivers, but in England, it flourishes in
very dry situations, and will bear exposure to
any degree of wind, thriving best when within
reach of the sea-breeze. The stems and leaves
contain a large quantity of sulphate of soda, a
fact which accounts for its flourishing not only
in such situations, but in the valleys of Arabia,
where the springs are often impregnated with salt.
It is held in high estimation in that country, for
its medicinal properties, which appear to have
recommended it also to the Romans, and indeed
to some of our own countrymen, among whom
was Archbishop Grindall.

The branches of the Arabian variety are commonly loaded with gall-nuts, which, before they
dry up, are full of a beautiful bright red sap, and
being exceedingly astringent, are collected and
used in dyeing. The people of Egypt generally
use the wood for fuel and building; bowls and
drinking-vessels are also made of it. The Arabs
cultivate it on account of the hardness of the
wood, which they use for the saddles of their
camels, and for other articles that are exposed to
rough handling. Burckhardt, in the account of
his journey through the wilderness of Sinai, says,
that it grows in great profusion in a valley to the
north of Mount Serbal, and that the Arabs
obtain from it a substance which they call *mann*,
and which closely resembles the description of
the manna given in Scripture. In the month

of June it drops from the branches upon the
fallen twigs and leaves, which always cover the
ground beneath the tree in its natural state.
The manna is collected before sunrise, when it
is coagulated, but it dissolves as soon as the sun
shines on it. The Arabs clear away the leaves
and dirt, which adhere to it, boil it, strain it
through a coarse piece of cloth, and put it into
leathern skins. In this way they preserve it till
the following year, and use it as they do honey,
to pour over their unleavened bread, or to dip
their bread into. I could not learn that they
ever made it into cakes or loaves. The manna is
found only in years when copious rains have
fallen; sometimes it is not produced at all. It
never acquires that degree of hardness which will
allow of its being pounded, as the Israelites are
said to have done with the manna with which
they were miraculously supplied, nor does it possess the same nutritive properties. Some travellers suppose this substance to be the produce of
an insect which infests the Tamarisk. The quantity collected is very trifling, perhaps not amounting to five or six hundred pounds, even in seasons
when the most copious rains fall. It is entirely
consumed among the Bedouins, who consider it the
greatest dainty which their country affords. The
harvest usually begins in June and lasts six weeks.

We may infer from this account, that although
the "bread from Heaven," supplied to the Israelites, and the manna of the Tamarisk, are as distinct from each other as any substances can be,
there was just enough outward resemblance between them to account for the name of manna
being given to their new food, supposing that the

mann of the Tamarisk was then known by the same name that it now is. On the other hand, it is highly probable, that the Arabs called the substance which they collected from the Tamarisk *mann*, from its bearing a resemblance in some respects to the manna of the Israelites. It is hard to say which of these opinions carries the greater weight; the supposition is quite natural, that the Israelites, amazed and perplexed at the suddenness of the miracle wrought on their behalf, called their new food by the name of the substance which it most resembled, and it is as natural that the Arabs would have given the name of manna to a white, sweet substance which they found on the ground before sunrise, although produced for a few weeks only in every year, and unaccompanied by the signs of a miraculous origin which characterized the food with which the Israelites were fed for forty years in the wilderness. But if, as Josephus tells us, the word manna means, "what is this?" and indicates ignorance of its nature and origin, there can be no doubt that the second opinion is the correct one. In no case is there any real connection between the two substances.

The manna of commerce, as has been stated before (vol. i. p. 146), is the produce of a European tree, *Ornus Europæa*.

Among the ruins of Babylon, on a spot which is supposed to be the site of the Hanging Gardens of Nebuchadnezzar, stands the famous solitary tree called by the Arabs, Atheleh; it bears every mark of antiquity in appearance, situation, and tradition. Its trunk was originally enormous, but, worn away by the lapse of ages, it is now a ruin amid ruins; nevertheless it bears spreading and evergreen branches, which are peculiarly beautiful, being adorned with long tress-like tendrils, resembling heron-feathers, growing from a central stem. These slender and delicate sprays bending towards the ground, give the whole the appearance of a Weeping-willow, while their gentle waving in the wind, whenever a breeze blows, produces a low and melancholy sound. Some travellers call it a Cedar; others say that it is a tree the like of which is not to be found elsewhere; finally, a very ancient tree, perhaps even as old as the time of Herodotus (B.C. 440). This tree, according to dried specimens gathered by Aucher in 1835, is the *Tamarix pycnocarpa*. Other species of Tamarisk grow in those countries; among others, *Tamarix corticulata*, as well as *Tamarix Gallica*, which last is found almost everywhere.

The Tamarisk was by the Greeks called Myrica; but the plant known to modern botanists by this name is a low shrub, composed of numerous upright stems, and producing in spring abundance of purplish brown catkins, which appear before the leaves begin to expand. It is commonly known as Dutch Myrtle or Sweet Gale, and to the latter name at least it is justly entitled, for both at the season when it is in flower, as well as when it is in leaf, it diffuses a rich aromatic perfume, which scents the air to a great distance.

> " And as he flies,
> Like the winged shaft, the wanton zephyrs breathe
> Delicious fragrance ; for upon his banks,
> Beautiful ever,—Nature's hand has thrown
> The odorous Myrica."   CARRINGTON.

The catkins and leaves, when bruised, are clammy

to the touch, and impart a permanent fragrance to the fingers.

DUTCH MYRTLE, OR SWEET GALE.

It is a native of Great Britain, North America, and all the colder and temperate regions of Europe

and Asia, always growing in bogs. The whole
plant abounds with a resinous substance, to
which it owes its fragrance. The leaves are
bitter, and are sometimes used as a substitute for
hops. The catkins when boiled throw up a
resinous wax, which may be made into candles.
This substance is found in much greater quantities in a North American species of Myrica, called
the Candlebury Myrtle. The plant which produces it is an evergreen, larger than the Sweet
Gale, and furnished with leaves like those of the
Sweet Bay. Candles formed of this wax burn
long, and yield a grateful smell, and they are said
to have the advantage of producing an agreeable
aromatic fragrance when extinguished. Another
species, which grows at the Cape of Good Hope,
produces a similar wax, which is applied to the
same purpose.

THE STRAWBERRY-TREE.

ARBUTUS UNEDO.

Natural Order—ERICACEÆ.

Class—DECANDRIA.   *Order*—MONOGYNIA.

THIS beautiful evergreen shrub, is better known by its ancient name of Arbutus,* than by the name which it derives from the fruit to which its berries bear a considerable resemblance. It is frequently mentioned by the Latin poets, as an ornamental tree, which added much grace to the wild rocky scenery of Italy, affording a shady retreat to the weary traveller, and food to the wild goat :—

> " Nunc viridi membra sub Arbuto
> Stratus, nunc ad aquæ lene caput sacræ."
> HORACE.
> " Now stretch'd beneath the Arbutus' green shade
> And now beside the sacred fountain laid."

> " Jubeo frondentia capris
> Arbuta sufficias."   VIRGIL.

> " With leafy Arbutus your goats supply."

We learn from Pliny that it was also called Unedo, or *One-I-eat*, the fruit not being nice enough to tempt any one to taste a second. He also notices the similarity between its fruit and

---

* The correct pronunciation of *Arbutus Unedo,* is with the accent on the first syllable of each word.

that of the strawberry, for he says that it is the only tree which bears fruit like ground-fruit. He also states, but not on his own authority, that in Arabia it attains an extraordinary height, evidently confounding it with some other tree.

The Arbutus is a native of the mountainous districts of Southern Europe and Northern Africa and of many parts of Asia. In England it only appears in the shrubbery and park. Among the rocky cliffs of Mount Edgecumbe in Devonshire, it flourishes in the immediate neighbourhood of the sea, but it never attains the dimensions of a tree. In Ireland it grows in great abundance about the hills and islands of Killarney; and here it is undoubtedly wild, though unfounded stories are told of its having been introduced by the monks of St. Finnian, in the sixth century. The country people in this neighbourhood eat the fruit, and Babington, whose judgment as a botanist few would call in question, pronounces it excellent. English berries, when thoroughly ripe, are of a mealy consistence, and of a somewhat insipid flavour, not unlike that of the haw. At Smyrna and Padua, it is exposed for sale in the markets, and the fruit which it produces near Miletus in Asia Minor, is said to resemble a strawberry, both in size and flavour. It is very probable, therefore, that when growing under certain conditions, the fruit improves in quality; indeed, Pliny intimates, that the produce of the tree varies, and Sir James Smith tells us, that in the Levant it is agreeable and wholesome.

The Arbutus is an evergreen shrub, with a scaly stem, and with dark green, glossy leaves, smaller than those of the Laurel, and serrated

at the edge. The flowers grow in clusters on
stalks bent downwards ; they are nearly globular
in shape, very elegant, of a greenish semi-trans-

FLOWER AND FRUIT OF ARBUTUS.

parent white, with a tinge of red. They expand
in September and October, and as they contain
a great deal of honey, are frequented by numbers

of the later butterflies and moths, wild bees, and wasps. The fruit, which takes a year to perfect itself, begins to ripen when the flowers expand; it is of the size of a cherry, and very like a strawberry, being covered with hard tubercles formed by the seeds, which are, however, not simply half-embedded in the berry, like the seeds of the strawberry, but concealed beneath the cuticle. It is most beautiful at the fall of the year, when its waxy flowers and scarlet berries present a very cheerful appearance.

A sugar, and a very good spirit have been extracted from the berry, and the leaves, it is said, may be employed with advantage in tanning. The wood is of little value, but at Killarney is manufactured into boxes and toys, which are sold to visitors; it is of a dull brown tint, and marked with fine lines, which are of a yet darker hue.

A variety is cultivated, which has red flowers, but it is scarcely more beautiful than the common kind.

The Arbutus gives as marked a character to the hills of Killarney, as the Box-tree does to the famous hill in Surrey, to which it has given the name of Box-hill. Mrs. S. C. Hall says: "The tourist, on approaching the lakes, is at once struck by the singularity and the variety of the foliage in the woods that clothe the hills by which on all sides they are surrounded. The effect produced is novel, striking, and beautiful; and is caused chiefly by the abundant mixture of the tree-shrub (*Arbutus Unedo*) with the Forest-trees. The Arbutus grows (not wild) in nearly all parts of Ireland; but nowhere is it

found of so large a size, or in such rich luxuriance, as at Killarney. The extreme western
position, the mild and humid atmosphere (for
in Ireland there is fact as well as fancy in the
poet's image—

> Thy suns with doubtful gleam
> Weep while they rise "),

and the rarity of frosts, contribute to its propagation, and nurture it to an enormous growth,
far surpassing that which it attains in any part
of Great Britain; although, even at Killarney,
it is never of so great a size as it is found
clothing the sides of Mount Athos. In Dinis
Island there is a tree, the stem of which is *seven
feet* in circumference, and its height is in proportion, being equal to that of an Ash-tree of
the same girth which stands near it. There are
several others nearly as large, and we believe
one or two still larger. Alone, its character is
not picturesque: the branches are bare, long,
gnarled and crooked, presenting in its wild state
a remarkable contrast to its trim-formed and
bush-like figure in our cultivated gardens. Mingled with other trees, however, it is exceedingly
beautiful; its bright green leaves happily mixing
with the light or dark drapery of its neighbours
—the Elm and the Ash, or the Holly and the
Yew, with which it is almost invariably intermixed. It strikes its roots apparently into the
very rocks—thus filling up spaces that would
otherwise be barren spots in the scenery. Its
beautiful berries, when arrived at maturity, are
no doubt conveyed by the birds who feed upon
them to the heights of inaccessible mountains,

II.   o

where they readily vegetate in situations almost
destitute of soil.   Its most remarkable peculiarity
is, that the flower (not unlike the lily of the
valley) and the fruit, ripe and unripe, are found
at the same time together, on the same tree.
The berry has an insipid though not an unpleasant taste, is nearly round, and resembles in
colour the wood-strawberry, whence its common
name, the Strawberry-tree.   It appears to the
greatest advantage in October, when it is covered
with a profusion of flowers in drooping clusters
and scarlet berries of the last year; and when
its grey green is strongly contrasted with the
brown and yellow tints which autumn has given
to its neighbours."

THE SPINDLE-TREE.

EUONYMUS EUROPÆUS.

*Natural Order—*CELASTRACEÆ.

*Class—*TETRANDRIA.   *Order—*MONOGYNIA.

FEW persons can have walked through a woodland district in September or October, without
noticing among the brush-wood, a straggling shrub
with remarkably green branches, narrow, smooth
leaves, and four-lobed seed-vessels, which split
vertically and disclose as many seeds, which are
wrapped up in a bright scarlet membrane.   This
is the Spindle-tree, a common shrub throughout
the whole of Europe, sometimes attaining a height
of from fifteen to twenty-five feet, but more generally ranking only as a hedge-bush.   Its flowers
appear in May : they are of four petals, small, and
of a whitish green colour.   The leaves and bark
are acrid and poisonous.   The wood, like that of
the Cornel, is of a very close grain, and being
used for the same purposes as that tree, is often
called by the same names, Prickwood and Dogwood.   It has long been used for making spindles,
whence it derives its name.   In Ireland it is commonly called Peg-wood, from its being made into
the pegs used by shoemakers.   Loudon says, that
it was formerly employed in the manufacture of
musical instruments, and that it is still occasionally used for the keys of pianofortes.   In
Scotland it is employed with the dark wood of

the Alpine Laburnum to form the drinking cups
called bickers. In making these, staves of the
yellow wood of the Spindle-tree, and of the dark

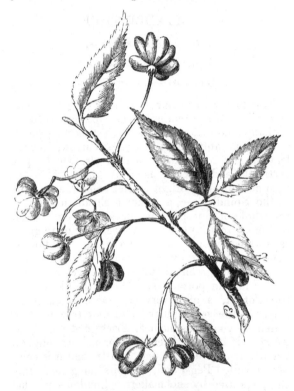

BRANCH OF THE SPINDLE TREE.

wood of the Laburnum are arranged alternately,
and produce a pleasing effect. In Germany,
spindles are still made of the wood, and in this

country watchmakers prefer it to any other kind
of wood for the slender spills which they use in
cleaning watches. When reduced to charcoal it
makes an excellent crayon for artists, being of
a strong texture, and making a mark which is
easily effaced. Loudon also states that the fruit
is sometimes employed by dyers, who derive a
yellow dye from the seeds boiled alone, a green
dye from the seeds boiled with alum, and a red
dye from the seed-vessels.

A variety of this tree is cultivated, which
bears scarlet seeds in a white seed-vessel. Several foreign species are also cultivated, all of
which, as well as the common one, are liable to
be entirely stripped of their foliage by the caterpillars of a moth, which cover the branches with
festoons of a web spun by them in the course
of their feeding.

THE DOGWOOD.

Cornus sanguinea.

Natural Order—Corneæ.

Class—Tetrandria.   *Order*—Monogynia.

LEAF AND FLOWER OF THE DOGWOOD.

This common hedge-shrub derives its Latin name from *cornu*, a horn, from the toughness of its wood; it is called Dogwood because "the fruit

is not fit even for a dog," on which account, also, it
was formerly named Dogberry and Hound's-tree.
It is also called Prickwood, from its wood being
frequently made into tooth-picks and skewers.

It usually grows in the form of a thick bush;
but may occasionally be seen trained up to be a

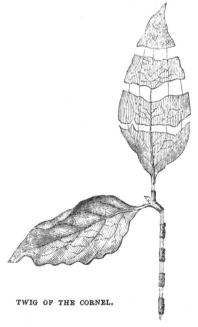

TWIG OF THE CORNEL.

round-headed tree, from fifteen to twenty feet
high, and with a stem six inches in diameter.
Unlike most other trees, it is best distinguished
in winter, when its bright red branches (which in
some places give it the name of Bloody-twig) are

very conspicuous. In early spring it bears at the extremities of the twigs numerous white flowers, which consist of four spreading petals, and these are succeeded by small berries, which in August and September become dark purple or almost black. Towards the end of September, the leaves turn bright red, and finally deep purple. The leaves and young shoots are remarkable for the number and strength of their spiral vessels. So tenacious are these, that if a tender twig or leaf (in any stage of growth) be snapped asunder in several places, the spiral vessels may be discovered by the naked eye, holding the pieces so firmly together that if one fragment be lifted up, all the others will remain suspended.

The fruit of the Dogwood is bitter and styptic, and yields an inferior kind of oil. The wood has long been used for the manufacture of small instruments, in which compactness of substance is required, such as arrows, ram-rods, &c., and it is still much sought after as a material for skewers.

Another species, *Cornus alba*, the white-fruited Dogwood, is common in shrubberies ; it resembles the last in habit, but its branches are longer and of a yet brighter red. It is a native of Siberia, and North America.

The Cornus, Cornel, or Cornelian Cherry of the ancients is another species, the *Cornus mas* of botanists. It grows wild in all the southern countries of Europe, except Great Britain, and may be distinguished from the preceding by its yellow flowers, ash-coloured twigs and scarlet berries, which are as large as acorns, but worthless as fruit. Mention of it occurs in several of the legends of the Romans, and its wood was much used for making spears. Our Dogwood appears to be the shrub alluded to by Pliny, under the name of *virga sanguinea*, bloody-twig.

THE PLANE.

THE ORIENTAL PLANE.

PLATANUS ORIENTALIS.

Natural Order—PLATANACEÆ.

Class—MONŒCIA.   *Order*—POLYANDRIA.

"TREES," says the Roman naturalist Pliny, "afforded the first inducement to the barbarous tribes of Gaul to cross the Alps, and spread themselves over Italy. A certain Swiss once came to Rome to learn the art of a smith, and on his return, took with him raisins, dried figs, oil and wine; the taste of which incited his countrymen to invade Italy with a hostile army. But who would have thought it possible that a tree should have been brought from a remote region of the world, for the sake of its shade only? Yet such was the case: the Plane was first carried across the Ionian Sea to shade the tomb of Diomede, who was buried in one of the small islands off the coast of Apulia; thence it was introduced into Sicily; from Sicily it was brought to Rhegium in Italy by the tyrant Dionysius; and has now extended so far, that the Morini (people of Calais) are taxed for its shade. Dionysius held it in high honour, and since his time it has so much increased in estimation, that its roots are nourished with wine instead of water."

Diomede was a Grecian hero, and to honour his tomb, the tree was planted which had of old been venerated in Greece, and even in Asia.

Herodotus informs us, that when Xerxes was
about to invade Europe with his mighty army,
and had arrived at Lydia in Asia Minor, he
fell in with a Plane-tree, which on account of
its excessive beauty, he decorated with golden
ornaments, and left behind him a warrior selected
from the Immortal Band to take care of it.
" Ælian and other authors tell us, " says
Evelyn, "he made halt, and stopped his prodigious
army of 170,000 soldiers, which even covered the
sea, exhausted rivers, and thrust Mount Athos
from the continent, to admire the pulchritude
and procerity of one of them; and became so
fond of it, that spoiling both himself and his
great persons of all their jewels, he covered it
with gold, gems, necklaces, scarfs and bracelets,
and infinite riches. In sum, he was so ena-
moured of it, that for some days neither the
concernment of his grand expedition, nor interest
of honour, nor the necessary motion of his por-
tentous army, could persuade him from it. He
styled it his mistress, his minion, his goddess;
and when he was forced to part from it, he
caused the figure of it to be stamped on a
medal of gold, which he continually wore about
him. Wherever they built their sumptuous and
magnificent colleges for the exercise of the youth
in gymnastics, and where the graver philosophers
also met to converse together and improve their
studies, they planted walks of Platans, to refresh
and shade them. The great Roman orators and
statesmen Cicero and Hortensius, would exchange now and then a turn at the bar, that
they might have the pleasure to step to their
villas and refresh their Platans, which they

would often do with wine, instead of water. And so prized was the very shade of this tree, that when afterwards they transplanted it into France, they exacted a *solarium*, by way of tribute, on any of the natives who should presume to put his head under it."

This veneration for the Plane still lingers in the East. The great Plane of the island Stanchio (anciently Cos) in the Archipelago, is remarkable for its size and the care with which the natives have attempted to preserve it. It has stood for time immemorial in the chief town of the island, and while it is the boast of the inhabitants, it is also, and with justice, the wonder of strangers. Earl Sandwich saw it in the year 1739, and calls it a Sycamore. "Among the curiosities of this city is a Sycamore-tree, which is, without doubt, the largest in the known world. It extends its branches, which are supported by many ancient pillars of porphyry, very antique, and other precious marble, in the exact form of a circle; from the outward verge of which, to the trunk, I measured forty-five large paces. Beneath the shade of this sycamore is a very beautiful fountain, round which the Turks have erected several chiosks, or summer-houses, to which they retire in the heat of the summer, and regale themselves with their afternoon coffee and pipe of tobacco." Dr. Clarke saw the same tree many years after: one enormous branch had then given way, notwithstanding its being supported by pillars of granite, and this loss considerably diminished its bulk. "Some notion," he says, "may be formed of the time those props have been so employed by the appearance of the bark; this has encased the extremities of the columns so completely that the branches and the pillars mutually support each other; and it is probable, if those branches were raised, some of them would lift the pillars from the earth." A specimen of this tree was given by Hasselquist to Linnæus, and it is now in the Linnæan Herbarium.

The Rev. S. Clark, of St. Mark's College, has furnished me with an interesting account of a Plane-tree, yet standing at Vostitza, a town on the Gulf of Lepanto, the ancient Corinthian Gulf.

"VOSTITZA, January 22nd 1844. — At this place (the ancient Ægium, where Agamemnon assembled the kings before the expedition to Troy, and which in later times was the head of the cities of the Achæan league) near where we landed, is an oriental Plane-tree of great beauty and vast size, said by the inhabitants to have stood from the time of Alexander. A more probable tradition fixes its age at eight centuries. We measured the trunk carefully, and found it forty-six feet in circumference. The branches spread nobly, but perhaps not quite in proportion to the height; the symmetry of the tree would else be as remarkable as its size. There is a large hollow in the trunk, which was formerly used as a prison, the opening being of moderate size, and easily fitted with a door. The tree stands in the lower town, at the foot of the somewhat tall cliff of loose conglomerate, on which the upper town is built. I suppose this situation has favoured its growth upwards.

"The Plane-tree was the tree employed by the Greeks to make a useful shade; and you find

them now in the public places in most towns. It
was in ancient times esteemed a good work to
plant Plane-trees in the Agora, or market-
place."

Buckingham describes this same tree as having
a trunk fifteen feet in diameter, and a hundred
feet in height, and as being covered with rich and
luxuriant foliage.

Just such another giant as this is described by
Pliny growing in Lycia. It stood near a fountain
by the road-side, and overshadowed a large area
with its tree-like branches. The trunk contained
a cave eighty feet in circumference, which was
set around with seats of moss-covered stones.
The Consul Lucinius Mutianus, when lieutenant
of the province, gave a banquet to seventeen of
his friends in this natural chamber, and all agreed
that the novelty of the scene gave a greater zest
to the viands than statues, pictures, or carved
work.

But the celebrity of the Plane-tree extends to
countries yet more remote than Asia Minor. In
Persia, where it is called Chenar, it has been
venerated from the earliest period, and according
to some, was formerly considered a protection
against the plague. The Persian gardens are
generally intersected by avenues of these trees,
and under them the inhabitants prefer to perform
their devotions. Sir William Ousely mentions
that the devotees sacrifice their old clothes by
hanging them to the branches; and that the trunks
of favourite Chenar-trees are commonly found
studded with rusty nails and tatters; the clothes
sacrificed being left nailed to the tree till they
drop to pieces of themselves. A similar custom,

it has been before remarked,* exists in Ireland, Arabia, and South America.

The Plane is a majestic tree with a massive smooth trunk. The bark is of an ash-grey, and is remarkable for pealing off in large thin flakes; so that the trunk does not borrow anything in size, like most other trees, from numerous deposits of bark. The leaves are large, and present a wide flat surface, from which circumstance the tree derives both its Greek and English names. The Oriental Plane is distinguished from the Occidental by having its leaves cut into five deep lobes, with numerous secondary notches, bearing a not altogether fanciful resemblance, pointed

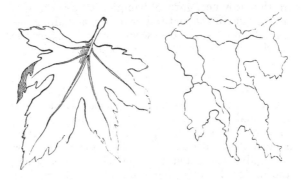

out by the ancient geographers, to the outline of the Morea, with its numerous bays and promontories. The flowers, which are produced in globular drooping heads, several together, appear with the leaves in spring, and are succeeded by balls of seed,

* Vol. i. pp. 182—184.

which are much smaller than those of the other
species, and remain attached to the tree all the
succeeding winter. By these the tree may be distinguished when every leaf has disappeared, as it
may also be by the light coloured irregular patches

LEAF AND FLOWER OF ORIENTAL PLANE.

on the trunk, occasioned by the shedding of the
bark described above. The seeds are imbedded
in soft bristly down, which, when the balls open,
serves to waft them away.

The Plane is now, as it was in Pliny's time,
only valued as an ornamental tree. Its wood is
smooth-grained, and susceptible of a high polish.
Loudon says, that it is not much used in the West
of Europe, but that in the Levant and Asia, it is employed in carpentry, joining, and cabinet-making.

II.   P

It is yellowish-white, till the tree attains a considerable size; after which it becomes brown, with jasper-like veins; and, wood of this kind, being rubbed with oil, and then highly-polished, resembles the wood of the Walnut.

It is a fast-growing tree, and prefers a good soil near running water, and is propagated, either by layers, cuttings, or seeds, the first method being preferred.

The largest existing Plane-tree is at the same time the largest tree in the world. It stands at Buyukdere, on the Bosphorus, and is described by various travellers. Dr. Walsh, who measured it in 1831, found the trunk *one hundred and forty-one feet in circumference* at the base, dividing into fourteen branches, some of which issue from below the present surface of the soil, and some do not divide till they rise seven or eight feet above it. One of the largest is hollowed out by fire, and affords a cabin to shelter a husbandman. De Candolle computes that it must be more than 2000 years old.

By the side of this monster all our English specimens dwindle into insignificance; indeed the only Oriental Plane existing in this country, to which any particular interest is attached, is one mentioned in Evelyn's Diary :—

" *September* 16*th*, 1683.—At the elegant villa and garden of Mr. Bohun at Lee. He shewed me the Zinnar-tree, or Platanus, and told me that since they had planted this kind of tree about the city of Ispahan in Persia, the plague, which formerly much infested the place, had exceedingly abated of its mortal effects, and rendered it very healthy." This tree is yet standing, and is exquisitely figured by Strutt in his Sylva. He says, "Lee Court remains at present much in the state in which it was during Evelyn's time; and the idea of this Plane-tree having been examined by him with curiosity and interest, as one of the first introduced into this country, is sufficient to give it value in the eyes of all who are acquainted with his admirable genius and virtues, independent of the attraction which it may boast on its own beauty. The circumference of this tree at six feet from the ground, is fourteen feet eight inches; it rises to the height of about sixty-five feet, and contains three hundred and one feet of timber."

Though the Plane was very rare in Evelyn's time, it was introduced a hundred years before, for Turner, in his Herbal (1541-1568), says:—

"I have seene two very young trees in England, which were called there Playn trees; whose leaves in all poyntes were lyke unto the leaves of the Italian Playn tre. And it is doubtles that these two trees were either brought out of Italy, or of som farr countre beyond Italy, whereunto the frieres, monks, and chanons went a pilgrimage."

OCCIDENTAL PLANE.

PLATANUS OCCIDENTALIS.

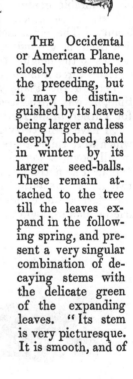

THE Occidental or American Plane, closely resembles the preceding, but it may be distinguished by its leaves being larger and less deeply lobed, and in winter by its larger seed-balls. These remain attached to the tree till the leaves expand in the following spring, and present a very singular combination of decaying stems with the delicate green of the expanding leaves. "Its stem is very picturesque. It is smooth, and of a light ash-colour, and has the property of throwing off its bark in scales; thus naturally cleansing itself, at least its larger boughs, from moss, and other parasitical encumbrances. This would be no recommendation of it in a picturesque light, if the removal of these encumbrances did not substitute as great a beauty in their room. These scales are very irregular, falling off sometimes in one part, and sometimes in another; and, as the under bark is, immediately after its excoriation, of a lighter hue than the upper, it offers to the pencil those smart touches which have so much effect in painting. These flakes, however, would be more beautiful if they fell off in a circular form, instead of a perpendicular one. They would correspond and unite better with the circular form of the bole. No tree forms a more pleasing shade than the Occidental Plane. It is full-leaved, and its leaf is large, smooth, and of a fine texture, and is seldom injured by insects. Its lower branches, shooting horizontally, soon take a direction to the ground, and the spray seems more sedulous than that of any tree we have, by twisting about in various forms, to fill up every little vacuity with shade. At the same time it must be owned, the twisting of its branches is a disadvantage to this tree, when it is stripped of its leaves and reduced to a skeleton. It has not the natural appearance which the spray of the Oak, and that of many other trees, discovers in winter. Nor indeed does its foliage, from the largeness of the leaf, and the mode of its growth, make the most picturesque appearance in summer."*

The leaf of the Plane exhibits one of those ex-

* Gilpin's " Forest Scenery."

quisite arrangements for the preservation of the bud, which confirms the exclamation of the poet:

" Each leaf and bud
Doth know I AM."

Trees, for the most part, produce new buds in the axil or angle between the leaf-stalk and

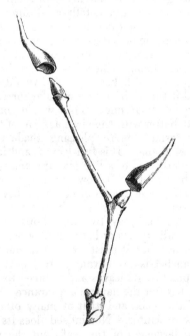

LEAF-STALK OF OCCIDENTAL PLANE.

the stem, and in many cases these buds attain a large size nearly a year before they expand.

The Plane appears to be an exception to this
rule ; for if the tree be examined while in full
foliage, not a single bud can be detected. On

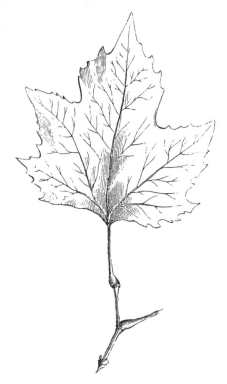

LEAF OF OCCIDENTAL PLANE.

a closer inspection, however, it will be observed,
that the leaf-stalk is much swollen at the base,
being there as thick as the twig which supports

it. Carefully detach it, and it will be found to be hollow, and to enclose a green bud, which remains behind after the leaf has been removed. The use of the hollow stem is to shelter the young bud during the colds of autumn. But when the leaf has fallen off, the bud is not left without protection, for it is enveloped in a tough case lined with a kind of resin, which is impenetrable to wet; within this is a similar case lined externally with the same coating; next come a number of scales covered with a dense coat of fur, which must serve as an admirable safeguard against cold; and within all lie the leaves, wrapped up in a mantle of silk, waiting till the succeeding spring shall give them new vigour and enable them to burst all their envelopes. For some time after their expansion, the leaves have their under surfaces covered with a thick coat of down, which circumstance has given to this Plane the name of Cotton-tree. Nor is this complicated protection against cold more than is requisite; for stout and lusty as the tree appears to be, it would without it be unable to withstand the insidious frosts of an English spring. In confirmation of this statement I may observe, that at the beginning of the present century large specimens of the Occidental Plane were not uncommon. In the January of 1809, however, there was a great flood, occasioned by a sudden thaw; and in the March and April following there was very mild weather, which tempted the Plane-trees to put on their leaves earlier than usual. This was succeeded by a severe frost in the beginning of May, which so injured the trees that they appeared sickly throughout all the summer;

and in the spring of 1810 a large number perished. The severe winter of 1813 destroyed a number of those which survived the frost of 1810, so that full-grown trees are now comparatively rare throughout Britain. Lofty trees may still be seen here and there with some of their branches dead or shivered by the tempest, the surviving boughs bearing scarcely enough leaves to enable us to distinguish the species, and affording a melancholy contrast to their ancient crown of foliage. Many persons suppose that this ruin is the effect of lightning, and have gone so far as to imagine that the Plane possesses some particular attraction for the electric fluid; but there can be little doubt that all these trees are among the sufferers from unseasonable frosts, that they have dwindled away under the effect of repeated shocks, and given up their dead and decaying boughs one by one to the violence of tempestuous winds. It does not appear that in its native country, North America, the Plane is injured by frost, although it is there exposed for a long period in every year to an intensity of cold unknown in Britain: hence it would appear that as long as the buds, the vitals of the tree, are protected by their many mantles, they defy the frost; but that if cold weather should return after the leaves have begun to expand, they become frost-bitten and perish.

In the swampy forests of America, it flourishes in unimpaired magnificence, and surpasses in size and height every tree with which it is associated. It often sends up a massive trunk seventy or eighty feet before it begins to branch, and then sends out huge arms, any one of which exceeds in dimensions the other trees which stand around. Michaux mentions one growing on a small island in the Ohio, which measured forty feet in circumference at five feet from the ground : and another on the right bank of the same river, which sent up a columnar mass of timber, forty-seven feet in circumference to the height of twenty feet before it began to branch.   His host offered to shew him others equally large, a few miles off from this last station.

In the Atlantic States of America the Plane is commonly known by the name of Buttonwood, from the resemblance between its seed-balls and old-fashioned buttons.   In other states it is called Water Beech, Sycamore or Cotton Tree. In some parts, where it is very abundant, the inhabitants regard it with dread, as they think that the down, which in summer detaches itself from the leaves and floats about in the air, has a tendency, when inhaled, to produce irritation of the lungs and finally consumption.

The timber of the Plane is of no great value, on account of its liability to warp ; it is, however, remarkable for having its concentric circles interrupted by bright medullary rays, and it will take a good polish.   It is only used for the commonest purposes.

"The Occidental Plane," says Lindley, "has been universally allowed to usurp the place of the Orientals in our plantations and parks.   Nine tenths of the Planes in the country belong to it ; but it is so tender that it is rare to see a really handsome specimen, the leaves being half killed by spring frosts, and the foliage, at the best, thin and bare, compared with that of the Oriental Plane."

Another species of Plane, called the Sycamore-leaved, *Platanus acerifolia*, is said to have been often mistaken by travellers in the East for the true Oriental Plane. Its seedballs are as large as those of the genuine kind; but the leaves are broader and less cut. It is sometimes named in this country the Spanish Plane, though it is not found in Spain either in a wild or cultivated state. Dr. Lindley is of opinion, that wherever the Occidental Plane is said to have become a fine tree in this country, it is the Sycamore-leaved species that is really intended; he consequently recommends that the custom of planting Occidental Planes should be discontinued.

THE BUCKTHORN.

RHAMNUS.

*Natural Order—*RHAMNEÆ.

*Class—*PENTANDRIA.　　　*Order—*MONOGYNIA.

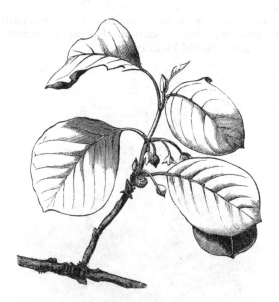

ALDER-BUCKTHORN.

Of this tree, little known though it probably
is, there are two very distinct species indigenous

to Britain, and one of them is far from uncommon. This species, *Rhamnus Frangula*, or Alder-Buckthorn, is a woodland shrub of upright growth, with a slender, purplish brown stem, and scanty, roundish leaves, which are remarkably smooth, both above and below, of a deep glossy green, entire at the edge, and conspicuously marked by many parallel veins. The flowers are green, small, and inconspicuous, and are succeeded by globular two-seeded berries, which, when ripe, are dark purple, and as large as peas. It is frequently to be met with in coppices, and among the undergrowth of woods, growing from six to ten feet high.

The other species, *Rhamnus catharticus*, resembles the last in habit, and grows in similar situations; it is, however, more bushy, and bears its flowers in clusters; the branches are more rigid; and the older branches are rough, and terminate in strong thorns. It may further be distinguished by its serrated leaves and four-celled berries, which are violently cathartic. This species attains a larger size in Siberia than with us. I have never observed it in the west of England, where the other species is common.

According to Loudon, "The juice of the unripe berries has the colour of saffron, and it is used for staining maps or paper; they are sold under the name of French berries. The juice of the ripe berries, mixed with alum, is the sap-green of painters; but if the berries be gathered late in autumn, the juice is purple. The bark affords a beautiful yellow dye. The tree does not make much show in spring, while in flower; but in autumn and winter, when profusely covered with black berries, it is very ornamental."

COMMON BUCKTHORN.

The Buckthorn possesses no little interest for the Entomologist, as it supplies food to the caterpillar of the Sulphur butterfly (*Gonepteryx*

Rhamni), which, the earliest of all its tribe, flutters about like a winged primrose on the first sunny days of March, and soon retires, to appear, however, once more, for a brief period, in September.

SULPHUR BUTTERFLY.

THE PRIVET.

LIGUSTRUM VULGARE.

Natural Order—OLEACEÆ.

Class—DIANDRIA.   *Order*—MONOGYNIA.

THE Privet would scarcely be considered entitled to a place among Forest Trees, were it not
for its being frequently associated in hedges with
the Hawthorn, a tree which requires no apologist.
Being, too, what may be called a half-evergreen,
it possesses a sort of claim on our notice.   In
its usual state it is a thick bush, with numerous
wiry stems, and smooth, somewhat fleshy, leaves,
shaped like those of the myrtle.   In June it bears
at the extremities of the shoots pyramidal clusters of white flowers, which have a sweet but
sickly odour, and soon change to a reddish-brown
colour.   The berries, which ripen in profusion,
are black, and remain attached to the tree until
almost every other kind of berry has disappeared,
when, as we may infer from their being left so
long untouched, they afford an unsavoury banquet to hungry birds.   In sheltered situations,
the leaves remain attached to the plant during all
the winter, but where the tree is much exposed,
it is stripped, at the fall of the year, of everything
but its clusters of dark fruit, which are sometimes
so numerous as to be very conspicuous.
     The Privet is commonly employed, either
alone, or in conjunction with the Hawthorn, in

the formation of hedges. Being very patient of the drip of trees, it is also often planted in shrubberies, to produce thickness of undergrowth, and,

FLOWER OF THE PRIVET.

from its indifference to the presence of coal-smoke, it is extensively used in the squares of large cities.

Q

Loudon states that a pink and a green dye may be prepared from the berries of the Privet, and that they also furnish, on pressure, a mild oil, which may be used for culinary purposes, as well as for lamps and making soap. In Belgium and Silesia the young twigs are used by the tanners; and the shoots are used, like those of the Osier, in basket-making, and for tying articles. From this last application, which was known to the Romans it acquired the name of Ligustrum, from *ligo*, to bind.

The caterpillars of several moths feed on the Privet; the most remarkable of these is the Privet Hawk-moth, *Sphynx ligustri;* it is among the largest found in England, in colour grass-green, with waved stripes of purple and white. The perfect insect is large and handsome.

THE PRIVET HAWK-MOTH.

THE ACACIA.

THE ACACIA.

ROBINIA PSEUD-ACACIA.

Natural Order—LEGUMINOSÆ.

Class—DIADELPHIA.   *Order*—DECANDRIA.

THE Acacia, or more properly, the False Acacia, though an American tree, is entitled to a distinct notice among British trees, from its having been one of the first trees introduced into England from North America, from its common occurrence, from the large size which it attains, and from the value of its timber. It was formerly supposed to be identical with the Acacia of Egypt, a tree which was imagined by the early missionaries to have supplied John the Baptist with food in the wilderness. From this circumstance it acquired the name of Locust-tree. The French called it Robinia, after M. Jean Robin (nurseryman to King Henry IV. of France), whose son was the first person who cultivated it in Europe.

The Acacia is a rough-barked tree, with straggling twisted branches, which are elegantly feathered with bright green, pinnated foliage. The flowers are shaped like those of the Laburnum, but larger, flesh-coloured or lilac, sweet scented, and hang down in dense clusters. The young shoots are smooth, of a purplish brown hue, and armed with rigid prickles. It was originally introduced and cultivated for its beauty, and during the summer months it is a most elegant

object. The principal objection alleged against it
is, that it is late in coming into leaf, and sheds its
foliage very early; the branches, too, being very
brittle, are liable to be snapped off by the wind,
even in situations which are not particularly exposed. Evelyn recommends it as "deserving a place
among our avenue trees, adorning our walks with
its exotic leaves and sweet flowers; very hardy
against the pinching winter; but not so proof
against its blustering winds." Gilpin says, that it
is often a very beautiful tree, whether it feathers
to the ground, as it sometimes does, or whether
it is adorned with a light foliage hanging from the
stem. But its beauty is very frail. It is of all
trees, the least able to endure the blast. In some
sheltered spot it may ornament a garden, but it
is by no means qualified to adorn a country. Its
wood is of so brittle a texture, especially when it
is encumbered with a weight of foliage, that you
can never depend upon its aid in filling up the
part you wish. The branch you admire to-day
may be demolished to-morrow. The misfortune is,
the Acacia is not one of those grand objects, like
the Oak, whose dignity is often increased by ruin.
It depends on its beauty rather than on its grandeur, which is a quality more liable to injury.

The Acacia grows with great rapidity when
young; seedlings often attain a height of from
twenty to forty feet in ten years, and established
young plants produce shoots eight or ten feet
long in one season. But when it has reached a
height of about forty or fifty feet, it grows very
slowly, and never acquires the dimensions of a
timber-tree. Loudon, who gives a long and elaborate account of the Acacia, attributes this pecu-

liarity to the fact, that its principal roots extend
themselves close to the surface of the ground,
where the soil is always richest. Hence the
growth is at first very rapid; but when the roots
cease to extend, all the surface-soil which they
traverse being exhausted, the growth of the tree
is retarded. But though it alters little in size
after it has reached its fiftieth or sixtieth year, it
is long-lived. The first tree that was introduced
into Europe by Monsieur Robin and planted in
the *Jardin des Plantes* at Paris, in 1635, is still
in existence, and is now seventy-five feet high.
About the year 1815 it shewed symptoms of decay, but the branches being lopped, the trunk has
shot out with redoubled vigour. This is, in all
probability, the oldest American tree in the eastern hemishere.

The wood of the Acacia is supposed to unite
the qualities of strength and durability to a degree
unknown in any other kind of timber; in consequence of which it has, for many years, been
employed throughout America and Europe in the
construction of the wooden pins, called trenails
(*tree-nails*), which are used to fasten the planks to
the ribs or timbers of ships. The history of the
first application of the Acacia to this purpose, is
given at length in Dr. Hunter's edition of Evelyn's Sylva, and is well worthy of repetition. It
is contained in a letter, dated July 25, 1782, from
Joseph Harrison, Esq., of Bawtry, to Dr. Hunter.
" The first experiment that I know of, respecting
the application of the timber of the Locust-tree
to any purpose in ship-building was in Virginia,
where I resided sometime about the year 1733:
and there, happening to be acquainted with an

ingenious shipwright, that had been sent over by
some merchants of Liverpool, to build two large
ships, I had frequent conversations with him
respecting the qualities of the several principal
timber-trees of that country. Being a person of
observation, he had made many useful remarks on
that subject, which the nature of his employment
afforded him many opportunities of doing with
advantage. He frequently spoke of the Locust-
tree as of extraordinary qualities both of strength
and duration; and used often to say, if a sufficient quantity could be had, it would be the best
timber he had ever met with for building ships.
After he had completed his engagements with his
employers at Liverpool, he set a small vessel
on the stocks for himself; but unluckily not
having a sufficient quantity of iron for the purpose, and none being to be had, at that time, in
the country, he was obliged to put a stop to the
work, till he bethought himself of the following
succedaneum. He had formerly (as hinted above)
observed the extraordinary strength and firmness
of the Locust-tree, and, on this emergency, took it
into his head that trenails of that timber might
be substituted for iron bolts,* in many places
where least liable to wrench, or twist, as in fastening the floor timbers to the keel, and the
knees,† to the end of the beams, which two articles take up a large proportion of the iron used

* BOLTS are round iron or copper pins, used to fasten the floor-
timbers to the keel, and the beams that support the decks, to the
sides of the ship, and on all other occasions where trenails are not
strong enough to bear the strain.

† KNEES are the crooked pieces of timber which are fastened to
the sides of vessels, and are intended to support the cross-beams.

in a ship, purposing, when he arrived in England
to bore out the Locust trenails, and drive iron
bolts in their stead. When he first informed me
of this scheme, I must own I thought the experiment very hazardous. However, as necessity has no
law, he put it in practice. The ship was built in
that manner, loaded, and sailed for Liverpool,
where she arrived safe; and though they met with
some blowing weather on the passage, she never
made so much water, but that one pump could
easily keep her free. She returned back to Virginia the next year, when I had an opportunity
of being informed by the builder himself (who
was then captain of her) what had been the result
of his project. He said, that during the passage,
especially in blowing weather, he was very attentive in examining the waterways,* as, at that
place, weak ships are most liable to work and
strain, but that he could not perceive any thing
more than is usual in other vessels. When un-
loaded she was hauled ashore upon the bank, in
order to be searched both outside and inside;
when, on the strictest examination, it was found
that the Locust trenails that had been substituted
for iron bolts, seemed (to all appearance) to have
effectually answered the purpose intended; however, it was thought prudent to take several of
them out and put in bolts in their room: and this
operation afforded another proof of their extraordinary strength and firmness, as they endure to
be backed out,† with a set-bolt, just as well as

* The WATERWAY is that part of a ship's deck that is next to the
sides of the ship; this seam, or joint, is very difficult to keep tight,
and in weak vessels will open and shut in carrying sail, when it blows,
hard.
† BACKING OUT a bolt or trenail is driving it out by means of a tool

though they had been iron; whereas Oak trenails, are usually bored out with an auger.   The next voyage the ship made was to the West Indies, where the captain died, and with him ended, for the present, any further prosecution of this matter; for, though the success of the above experiment was known to many, yet (as is frequently the case with new discoveries) none, that I ever heard of, made any use of Locust trenails in shipbuilding till many years after; though on the goodness of that article greatly depends the strength and durableness of a ship.   I frequently recommended it, when opportunities offered, but all to no purpose, till about twenty years ago, when I was settled in trade at Rhode Island, I persuaded some ship-builders to try the experiment; but notwithstanding all my endeavours, the use of Locust trenails still continued to be little practised or known, till it happened to be adopted by a builder of some eminence at New York, and of late years has been introduced into general use there, and in some parts of New England.   But, as yet, the use of the Locust-tree in ship-building is confined to the article of trenails, on account of its scarcity; for, were it nearly as plentiful as Oak, it would be applied to more purposes, being much superior to it, both as to strength and duration; and from its spreading into branches, affords full as large a proportion of crooks as that timber."

called a *set-bolt,* which is an iron punch, something smaller than the bolt or trenail to be taken out, against which it is driven with a heavy blacksmith's sledge-hammer; but Oak trenails, except such as are very hard and sound, will seldom bear this operation, in which case they are obliged to bore them out with an auger.

Mr. Harrison's efforts have been at length successful; the American shipwrights, Loudon tells us, now use as much Locust-wood as they can procure, finding it as durable as the Live Oak, and the Red Cedar, with the advantage of being stronger than the former and lighter than the latter. It is difficult, however, to procure trees of sufficient size for ship-building; for even, in those districts where the tree thrives best, nine-tenths of the trunks do not exceed one foot in diameter, and from thirty to forty feet in height. The wood is used for trenails in all the sea-ports of the Middle States, to the exclusion of every other kind of timber. Instead of decaying, it acquires an extraordinary degree of hardness with time. In 1819, from 50,000 to 100,000 trenails of this wood were exported to England, and their excellence has been confirmed by the highest authorities, so that Oak-wood grown in Sussex, which was formerly considered the best for this purpose, is now only partially used, Locust trenails being still imported from America to a very great extent.

It may seem strange that the timber of a tree so liable to be broken by the wind, as we find the Acacia to be, should be considered the very best for a purpose where extraordinary strength is required; but it must be remembered that the Acacia, in its native country, prefers barren, sandy, or light soils, in which situations it matures its timber slowly. With us, on the other hand, it is usually planted in rich soils and sheltered situations, where, though the tree is botanically the same, the character of the timber is materially impaired.

In the year 1823, Cobbett drew the public attention to the Locust-tree (then scarcely known
by that name), and recommended that it should
be extensively planted in England, for the sake
of its timber, which he asserted to be superior to
anything else for a variety of purposes, and
predicted that " the time would come when the
Locust-tree would be more common in England
than the Oak."   To supply the demand which he
had himself created, he imported enormous quantities of seeds from America, turned his garden
into a nursery, and "sold altogether more than a
million of plants."   But still, not being able to
raise enough plants to supply all his customers,
he purchased large numbers from the London
nursery-men, and fortunate did the applicant
consider himself, who could purchase at a high
price from Mr. Cobbett, the very same Locust-
trees that, under the name of Robinia Pseud-
Acacia, were standing unasked-for in the nurseries.
We are undoubtedly indebted to Mr. Cobbett
for very many of the Acacias that now adorn our
parks and pleasure-grounds; but it is far from
proved that the ground, which has been devoted to
plantations of these trees, might not have been
more profitably employed.   Loudon has shewn
satisfactorily that Cobbett's recommendation of
the wood cannot, in many instances, be confirmed
by fact, but allows that sound Acacia wood is
heavier, harder, stronger, tougher, more rigid, and
more elastic, than that of the best English Oak;
and, consequently, that it is more fit than Oak
for trenails."   He adds, moreover, that "it is
very suitable for posts and fencing, and also for
the axle-trees of timber-carriages; but that there

is no evidence of its being applicable to the
general purposes of construction."

Other parts of the tree are not without use;
the roots are very sweet, and afford an extract
like that obtained from liquorice root; and the
foliage forms an excellent substitute for clover as
provender for cattle.

There are many fine specimens of the Acacia
growing in England, varying in height from fifty
to eighty feet, and from six to ten feet in circumference; but none of these demand particular
notice.

THE HUNTINGDON WILLOW.

THE WILLOW.

SALIX.

Natural Order—AMENTACEÆ.

Class—DIŒCIA.   *Order*—TRIANDRIA.

BY the common consent of mankind, trees have in all ages been selected as affording the most appropriate emblems of the passions by which both states and individuals have been swayed, as well as to indicate the various changes in condition to which, from time to time, they have been subjected. I need only mention the Palm, the Olive, the Bay, the Cypress, and I recall at once the ideas of rejoicing, peace, victory, and mourning. The Willow is remarkable among these for having been in different ages emblematic of two directly opposite feelings, at one time being associated with the Palm, at another with the Cypress. The earliest mention of the Willow which occurs in any composition is to be found in the Pentateuch,* where the Israelites were directed at the institution of the feast of Tabernacles to " take the boughs of goodly trees, branches of Palm-trees, and the boughs of thick trees, and Willows of the brook, and to rejoice before the Lord their God seven days."

To wanderers in a dry and barren wilderness, the bare mention of a tree bearing the name of the " Willow of the Brook," must have come as-

---

* Lev. xxiii. 40.

sociated with the most pleasurable feelings; and
even when the Israelites were settled in a land
which was "the joy of all lands," this tree still
continued to be emblematical of joyful prosperity.
The prophet Isaiah, foretelling the glorious resto-
ration of Israel, says, "They shall spring up as
among the grass, as Willows by the water-
courses."* But while the Jews were in captivity
in a strange land, an incident occurred which,
to that nation at least, made the Willow an
emblem of the deepest of sorrows, namely, sorrow
for sin found out and visited with its due punish-
ment. "By the rivers of Babylon, there we sat
down, yea, we wept, when we remembered Zion.
We hanged our harps upon the Willows in the
midst thereof. For there they that carried us
away captive required of us a song; and they that
wasted us required of us mirth."

From that time the Willow appears never again
to have been associated with feelings of gladness.
Even among heathen nations, for what reason we
know not, it was a tree of evil omen, and was employed to make the torches carried at funerals.
Our own poets have made the Willow the symbol
of despairing woe; Spenser makes it the fitting
garb of the forlorn; Shakspeare represents the
doomed Queen of Carthage standing—

> "With a Willow in her hand
> Upon the wild sea banks;"

and Herrick says,

> "As beasts unto the altars go,
> With garlands dressed, so I
> Will, with my Willow wreath, also
> Come forth and sweetly die."

---

* Isaiah xliv. 4.

These poets, it should be remembered, wrote before the Weeping or Babylonian Willow was known in Europe; but there can be no doubt that the dedication of the tree to sorrow is to be traced to the pathetic passage in the Psalms, quoted above.

Few persons are aware how very large a number of species belong to the genus Willow. More than two hundred are described by Loudon, which are to be found growing in British collections; of these seventy are enumerated by Sir W. J. Hooker, as natives of Britain; Babington has reduced this number to fifty-seven, and Lindley, following the arrangement of Koch, has further reduced it to thirty; the two last authors considering as mere varieties some which were considered to be distinct species. This simple fact is enough to shew that the genus is a very difficult one even to botanists. In a work of this kind an attempt to give even a sketch of the principal species would be quite out of place, and it will perhaps be considered presumptuous if I give an opinion as to which of the three authors named above is most worthy of being selected as a guide. Nevertheless, I may state that Sir W. J. Hooker's " British Flora" contains an account of the species as arranged by Borrer, who is, I believe, allowed to be the first of living British botanists. Babington confesses himself to be imperfectly acquainted with many of those which Borrer considers species, but which he makes varieties; and Lindley, in the first edition of his " Synopsis," follows Sir J. Smith, enumerating sixty-four species, but in the second edition follows Koch, and reduces the sixty-four to thirty. I am much afraid that

this statement will not tempt many of my readers
to study the Willow-tribe botanically; but if it
does, they can select their own guide.

If modern science has done so little towards
reducing this unruly tribe to order, we must not
expect much accuracy from the older naturalists.
Accordingly we find that Pliny mentions only eight
species, and it cannot now be ascertained what
these were, for he distinguishes them more by the
names which they bore in his time, than by description. He places them among the most useful
of aquatic trees, furnishing vine-props, cordage,
osiers for fine and coarse basket-work, and rural
implements of many kinds. No tree, he says,
affords a safer return to the planter, gives less
trouble, or is more independent of the seasons. On
the authority of Cato, he assigns to it the third
rank among the most valuable of vegetable productions, placing it before Olive-yards, corn, and
pasturage.

The Willows are natives of the temperate regions
of the northern hemisphere, and are much more
numerous in the Old World than in the New. The
majority grow by the sides of water-courses, but a
few grow high up in the mountains, and are found
nearer to the North Pole than any other shrubby
plants. As far as it is possible to include under
a general description so extensive an array of
species, they may be characterized as trees or
shrubs varying in height from sixty feet to a few
inches. They grow rapidly, and readily shoot
from cuttings; the wood is white, the bark of the
trunk rather smooth than otherwise; that of the
branches downy or smooth, in the latter case sometimes to such a degree as to appear varnished. In

most species it is stringy and unusually tough, and in all is of a bitter taste, owing to the presence of a chemical principle called *salicine*, which possesses nearly the same medicinal properties as quinine, the substance which is extracted from Peruvian bark. The leaves are undivided, either notched at the edges, or even, stalked, often fur-

BLOSSOM OF THE CRACK WILLOW.

nished with stipules, smooth or silky, downy, or even cottony, and varying in shape from linear to round, some modification of the ellipse being, however, by far the commonest form. The flowers, which are catkins, appear early, and are of two kinds, each growing on a separate tree. The barren catkin, is an erect stem, closely invested on all sides with leafy over-lapping scales, within each of which are from two to five delicate stamens, with two-lobed yellow anthers (*fig. n.*) and a gland containing honey. Before expansion the catkin resembles a large silky bud, and is afterwards more or less oblong in shape. The catkin of the fertile tree is nearly the same as the barren catkin, but each scale contains, instead of stamens, a single pistil with two stigmas, which as it enlarges becomes an egg-shaped germen (*fig. o.*) of one cell and two valves. The seed-vessel, when ripe splits on its two opposite sides, the valves roll back (*fig. p.*) and disclose numerous minute seeds, each of which is tufted with downy or silky hair (*fig. r*). Some species of Willow are in full flower by the third week in March, and whenever a bright, warm day occurs after this time, the bees sally forth and resort in swarms to the fragrant catkins for a spring breakfast. I have noticed them busily engaged as early as the 22nd, and others have observed them yet earlier. The value of Willow-bushes near hives can scarcely be overstated, for this is just the season when the combs are likely to be exhausted, and there are as yet few other flowers in bloom capable of affording a considerable supply.

Before the discovery of sugar, honey was far more valuable than it is at present, and it appears from Virgil that, in his time, Willows were commonly planted in apiaries, for the special purpose of affording nourishment to the bees at this critical season:

Hinc, tibi quæ semper vicino ab limite sæpes,
Hyblæis apibus florem depasta salicti,
Sæpe levi somnum suadebit inire susurro.—*Ec.* i.

"The Willow-hedge, which parts your neighbour's land,
To bees of Hybla yields unfailing store
Of sweetest nectar, and with constant hum
Invites repose."

Owing perhaps to the association of the Willow with the Palm, in the passage quoted from Leviticus, blossoms of Willow, under the name of "Palms," are in some parts of Great Britain worn on the day which commemorates our blessed Lord's triumphal entry into Jerusalem. At Lanark, according to ancient usage, the boys of the Grammar-school parade the streets on the day before, carrying a Willow-tree in blossom ornamented with daffodils and other spring flowers. A writer in the " Every-day Book" says,—" It is still customary with men and boys to go a palming in London early on Palm Sunday morning; that is, by gathering branches of the Willow, with their grey,  shining, velvet-looking buds, from those trees in the vicinity of the metropolis; they come home with slips in their hats, and sticking in the breast button-holes of their coats, and a

WILLOW.

sprig in the mouth, bearing the Palm-branches in their hands. This usage remains among the ignorant from poor neighbourhoods, and there is still to be found a basket-woman or two at Covent Garden, and in the chief markets, with this ' Palm' as they call it, on the Saturday before Palm Sunday, which they sell to those who are willing to buy; but the demand of late years has been very little, and hence the quantity on sale is very small. Nine out of ten among the purchasers buy it in imitation of others, they care not why; and such purchasers, being Londoners, do not even know the tree which produces it, but imagine it to be a real Palm-tree, and ' wonder they never saw any Palm-trees, and where they grow.' "

The Willow ripens its seeds early enough to furnish many of the feathered tribe with a soft and warm material for lining their nests; and this too is all the more valuable from the fact that no other downy seeds are as yet ripe, and that the rains of winter have beaten into the earth all the thistle-down that had not been dispersed by the preceding equinoctial gales. In fine weather the air is often filled with the floating seeds, as they are wafted away to some suitable place of growth.* Loudon says, that this down is sometimes collected and used as a substitute for cotton in stuffing mattresses, and that in Germany a coarse kind of paper is made of it.

The leaves of several kinds of Willow are, on

---

* A part of the kitchen-garden at Versailles having been neglected during the first Revolution, and for many years after, indeed until 1819, the light downy seeds from Poplars and Willows in the neighbouring woods sprung up, and converted the whole place into a wood of timber-trees.

the continent, used as fodder for cattle, being
collected in summer and stacked for winter consumption. In Sweden and Norway the bark is
kiln-dried in seasons of scarcity, and mixed with
oatmeal. In the same countries the twigs are
twisted into ropes, as they were in Pliny's time,
which are used even for the cordage of vessels.
The inner bark is applied to the same purposes
as that of the Lime, and in Tartary is woven into
a coarse cloth. The wood is soft, smooth, and
light, and is applied to a great variety of purposes,
especially for fast-sailing sloops of war, and cricket
bats. Split into thin strips it is manufactured into
hats. The boats used by the early Britains were
constructed of Willow-rods, covered with hides;
they were called coracles, and it is curious that
very similar vessels, called by the Irish *currack*,
are in partial use to this day. " Coracles thus
made," says Southey, "and differing only in the
material with which they are coated, and carrying
only a single person, are still used upon the
Severn, and in most of the Welsh rivers. They
are so small and light, that when the fisherman
lands he takes his boat out of the water, and bears
it home upon his back." Boats of this description
were in common use on the Euphrates in the
time of Herodotus, B.C. 444. He says that the
Armenians, who carried on a traffic with Babylon,
built their boats of Willow,* covering the outsides
with skins, making them circular like a shield,
without distinguishing the prow from the stern.
Having placed their merchandize, principally Palm-
wine, on board, they cover it with straw and float
down the stream. The crew consists of two men,

† In Greek, ἰτέη, our *withy*.

who guide the vessel by oars. Each boat contains, besides the goods and rowers, a living ass, or if the vessel be a large one, several. On their arrival at Babylon they dispose of their merchandize, take their vessels to pieces, sell the Willow-ribs and straw, and having laden their asses with the skins, return home by land, the current not allowing them to sail up the stream. On some of the rivers of India, boats of a precisely similar form are used at the present time, some of them large enough to transport heavy artillery. The only difference appears to be that Bamboo is now used to form the ribs instead of Willow.

Pliny, quoting a more ancient author, says that the Britons used to make voyages to an island called Mictis, distant six days' sail, in vessels of the same construction as those described above, and to return with cargoes of tin. Julius Cæsar relates, in his History of the Civil War, that his recollection of the coracles which he had seen during his invasion of Britain, was, on one occasion the means of extricating his army from a critical position; for, being hemmed in by the enemy, and being unable to throw a bridge across a river which impeded his movements, he set his troops to work, and quickly completed enough boats to transport his army.

In a picturesque point of view, the Willows do not rank high; they are formal in their mode of growth, and are loaded with bundles of twigs, rather than with ramified branches: the foliage too is meagre, and is not disposed to form pleasing tufts. Gilpin does not recommend their use in artificial landscape, "except as pollards, to characterize a marshy country; or to mark in a

second distance the winding banks of a heavy,
low-sunk river, which could not otherwise be
noticed. Some Willows, indeed, I have thought

FOLIAGE OF HUNTINGDON WILLOW.

beautiful, and fit to appear in the decoration of
any rural scene. The kind I have most admired
has a small narrow leaf, and wears a pleasant,
light, sea-green tint, which mixes agreeably with

foliage of a deeper hue. I am not acquainted
with the botanical name of this species, but I
believe the botanists call it *Salix alba.*" This
is the Huntingdon, or White Willow, a good
specimen of which is figured at the head of this
chapter ; it derives its name from the silky whiteness of the under side of the leaf.

This species is said to be one of the most useful
of the genus as a timber-tree ; like the rest of the
Willows, it grows rapidly, and acquires considerable magnitude within the usual period of
human life, and may therefore, in the natural
course of events, be cut down, a full-grown tree,
by the same hand that planted it. " It groweth
incredibly fast," says Fuller, " it being a by-word
in this county, that the profit by Willows will
buy the owner a horse, before that by other trees
will pay for his saddle." The wood is soft but
elastic, and is well adapted for the lining of
barges and carts, which are used for carrying heavy
loads of hard substances. It is durable, and makes
good roofing. The bark is used by tanners, and
it makes excellent firewood ; added to which, it
grows without trouble from cuttings, and thrives
in any soil except peat, in which situation only
the smaller species will grow.

The Bedford Willow, *Salix Russelliana,* is
another of the tribe which attains a large size. It
was named in honour of the late Francis, Duke of
Bedford, by whom it was first brought into notice.
Its leaves are in shape very like those of the
White Willow, but differ in being smooth on
both sides. The favourite tree of Dr. Johnson
was of this species. It stood near the public
footpath in the fields, between the city and

Stow Hill, and the Doctor, in spite of his admiration of brick walls, used frequently to rest
under its shade, and give himself up to the sweet
influences of nature. The circumference of this
tree in 1781, was above fifteen feet; the trunk
rose to the height of twelve feet, and then
divided into fifteen large ascending branches,
which spread at the top like an Oak. The circumference of the branches was upwards of two
hundred, and it covered an area of nearly four
thousand feet. Its height was forty-nine feet.
In 1810 it stood in unimpaired vigour, having
increased to twenty-one feet in circumference,
the trunk ascending to a height of twenty feet,
before it branched; but in the November of that
year many of the branches were swept away by a
violent storm, and nearly half of what remained
fell to the ground in August 1815, leaving little
more than its stupendous trunk and a few side
boughs. Finally, in April 1829, it was blown
down, and its remains converted into snuff-
boxes, and similar mementoes of the great man
after whom it was named.

The timber of the Bedford Willow is said by
Loudon to be more valuable than that of any
other species; the bark contains more of the tanning principle than the Oak. It is in this species
also that salicine is most abundant.

The Crack Willow, *Salix fragilis*, derives its
name from the brittleness of the branches, which
start from the trunk under the slightest blow. Its
leaves closely resemble those of the Bedford Willow, but, according to Selby, " the ramification is
more oblique, and the branches in consequence
cross each other more. It is also less beautiful and

imposing in appearance, and seldom attains
so great a size. It is very subject to become
naked or stag-headed, by the decay of its up-
permost branches, though it continues to live and
throw out long annual shoots for many years
afterwards." When first cut, the sap-wood of
the Crack Willow is white, the heart-wood pale
red; upon exposure to the air, and when seasoned,
both become of a fine salmon colour. The roots
afford a purple red dye, and are still used in
Sweden and in France, to colour Easter eggs.

The Goat Willow, *Salix caprea,* is the common
Coppice and Hedge Willow, which affords so early
a banquet to the bee. It may readily be distinguished by its purplish brown branches, which
are covered with minute down when young, and
by its large broad leaves, which are wavy at the
edge and densely clothed beneath with soft, white,
cottony down.

This species never attains a large size, but it
forms a valuable underwood. In Scotland, where
it is called the Saugh-tree, the wood is said to
be much used for the handles of hatchets, rake-
teeth, and other articles which require to be
made of a light and tough material. The
branches also, when two or three years old, make
good hoops for casks.

The species of Salix which are used for
basket-making are usually called Osiers. Several
kinds are in common cultivation, all agreeing in
bearing long, flexible, tough shoots, and narrow,
pointed leaves. These being planted for the sake
of their young rods, are rarely suffered to attain
their full size, with the exception of the Golden
Osier, which is often to be found in gardens and

shrubberies, where its bright yellow branches are
very ornamental, especially in winter.

Those which are best adapted for basket-making
are the Common Osier, *S. viminalis*, and the Three-
stamened Osier, *S. triandra*. They should be
planted in low, and naturally moist situations,
and in a deep, well-drained soil, which, to be productive, should be kept well cleared of weeds. In
the second autumn after planting, the shoots are
fit to be cut, and the process is repeated every
year, immediately after the fall of the leaf, when
the wood is thoroughly ripe. If they are not
wanted to be used with the bark on, they are
tied up in bundles, and placed on end in standing
water until the following spring. When the buds
begin to shoot, the rods are ready for peeling,
and after this process they will keep for a very
long time. Of late years large quantities of
Osiers have been imported from Holland, in consequence of which Willow-holts in England
are far less profitable than they used to be.

Osiers are not unfrequently planted by the
way-side and in low meadows, as pollards, for the
purpose of supplying poles and stakes. The centres
of these trees very soon decay, and the young
buds send down roots into the mass of rotten
wood, sometimes until the cavity is nearly filled.
Dr. Plot, in his Natural History of Oxfordshire,
mentions some Pollard Willows, on which seeds
of Ash had been accidentally lodged and germi-
nated, so that " the roots of the Ashes had, some
of them, grown down through the whole length of
the trunks of the Willows, and at last fastening
into the earth itself, so extended themselves that
they burst the Willows in sunder, whose sides

falling away from them, and perishing by degrees, what before were but the roots, are now become the bodies of the Ashes themselves." Loudon records a yet more remarkable incident connected with this tree. An old Willow at Carlsruhe having been nearly thrown down by a storm, was supported by an oaken prop. Into this it sent down a root, which fixed itself in the earth, and as it increased in size, split off the bark from the prop, and eventually became so stout as to render the artificial support no longer necessary.

A beautiful species of Willow, which is not so generally cultivated as it deserves to be, is the Five-stamened Willow, *Salix pentandra*. This is easily distinguished by its large glossy leaves, more like those of the Portugal Laurel than of the other Willows. During the whole of summer it has quite the appearance of an evergreen, and towards the end of June is very conspicuous with its seed-vessels, which, being then ripe, burst, and disclose a great abundance of glossy silk attached to the seeds. It has this further recommendation, that the foliage emits a fragrant aromatic perfume. It grows in watery places in the north of England and Ireland. In the latter country I have seen bushy hedges of it stretching across the extensive bogs, which abound in the neighbourhood of the Giant's Causeway, scenting the air, and giving a cheerful appearance to the otherwise gloomy landscape. It forms a more compact tree than the other kinds, but the shoots are too brittle to be of much value.

The little Willow, which in some districts is so abundant on commons, trailing its wiry branches along the ground, or occasionally availing itself of the support afforded by Heath and Furze, to, assume the form of a dwarf shrub, is the Brown Willow, *Salix fusca*. Its leaves are glossy above, and very silky beneath; and its long twigs are conspicuous in May and June, from the numerous yellow catkins arranged at regular intervals along opposite sides of the stem. During the latter part of summer, the seeds of the fertile plant give to the ground the appearance of having been strewed with cotton.

Last and least among the British trees of this family comes the Herbaceous Willow, *Salix herbacea*. The ordinary height of this diminutive

HERBACEOUS WILLOW.

tree is about four inches. It is a native of many parts of Europe, and of North America, and in Great Britain is the last plant furnished with a woody stem, which we meet in ascending the mountains.* "In Switzerland," De Candolle observes, "some species of Willow spread over the uneven surface of the soil; and as their branches

* See "Botanical Rambles," page 123, 16mo edition.

are often covered with the earth, which the heavy rains wash over them, they present the singular phenomenon of trees which are more or less subterranean. The extremities of these branches form sometimes a kind of turf, and the astonished traveller finds himself, as we may say, walking on the top of a tree. *Salix herbacea* is the species that most frequently presents this remarkable appearance, because it generally grows on steep slopes of loose soil." The leaves are employed in Iceland in the tanning of leather.

A long list of insects that feed on the Willow is given by Loudon and Selby. Among them the most destructive is the larva or caterpillar of the Goat-moth.* The perfect insect lays, it is said, as many as a thousand eggs at the base of a tree, and the larvæ, as soon as they are hatched penetrate the bark, and spend three years of their existence in eating out long galleries in the wood. At first they are exceedingly small, but when full-grown are four inches long, and about an inch in circumference. They are then disgusting insects: the body is brownish-red above, yellow beneath, and sprinkled with hairs. When touched the caterpillar discharges from the mouth a fetid acrid fluid, which it is said to use in softening the wood on which it feeds. Before changing to a chrysalis it makes a strong cocoon of chips of wood, and small pieces of bark which it has gnawed off, and awaits its final transformation. The cocoon is placed at the end of the excavation, and in contact with the air, so that the perfect insect has no difficulty in effecting its escape. The moth measures above three inches from the

* *Cossus ligniperda.*

tip of one wing to that of the other. Its prevailing colour is grey, crossed with countless brown lines, of which two broader stripes towards the outer border are particularly conspicuous.

As a great number of caterpillars frequently attack the same tree, the ravages which they commit are very considerable. Not only is the timber spoiled, but in the course of a few years so much of the trunk is destroyed that the tree is very liable to be blown down, even in a moderate gale. The Elm, Walnut, and other trees are also subject to the attacks of these mischievous insects. The reader is referred to the authors above mentioned for a description of other insects, which, in the larva state, feed on the Willow.

The Willow is liable to the attacks of a Gall-fly, which, in the summer, lays its eggs in the young twigs. The effect is that the juices of the tree, diverted from their natural use, harden into an irregular mass, which, when

WILLOW GALL.

the young grubs are hatched, serves them both for food and dwelling. While the trees are in leaf, these galls are generally hidden from sight; but in winter they are often to be seen on the extremities of the branches, each containing a number of small white larvæ. They are of the same colour as the bark, and of a corky consistence; but when once formed, they have no vegetative power, the twigs which seemingly pass through them being always withered and dead.

Willows are common in the East, and are frequently mentioned in the Bible, as in the passages already quoted; in the Book of Job xl. 22, where Behemoth is said to be compassed about with Willows of the brook. Ezekiel (xvii. 5) in his figurative description of the last branch of the house of Judah, says, that a great eagle cropped off the topmost twig of a Cedar-tree, and set it by great waters as a Willow-tree.

Rauwolf states, that near "Halepo (Aleppo), about the rivulets, there is a peculiar sort of Willow-trees called Saf-caf, &c.; these are not all alike in bigness and height, and in their stems and twigs they are not very unlike to Birch-trees (which are long, thin, weak, and of a pale yellow colour); they have soft ash-coloured leaves, or rather like unto the leaves of the Poplar-tree, and on their twigs here and there are shoots of a span long, like unto those of the Cypriotish wild Fig-tree, which put forth, in the spring, tender and woolly flowers like unto the blossoms of the Poplar-tree, only they are of a more drying quality, of a pale colour, and a fragrant smell. The inhabitants pull of these (because they bear no fruits) in great quantities, and distil a very precious and sweet

water out of them very comfortable and corrobo-
rative to the heart."

Hasselquist, in reply to some questions asked by
a friend, says:

" Calaf is a little Willow which never grows to
a large tree; it has a straight trunk with a
smooth, oval, lancet-shaped leaf, deeply sawed on
the edges. No tree in Egypt is more famous
amongst the inhabitants, on account of the water
that is distilled in the spring from its blossoms,
which is much more used as a family medicine by
the Egyptians than treacle by our peasants.
They are scarcely afflicted with any disease, but
they use the water of Calaf. There are apothe-
caries in Cairo, whose chief, almost only, employment is to sell Calaf, for thus they likewise call the
water. It is cooling, promotes perspiration, and
is somewhat cordial; it therefore serves in the
continual fevers, which are so common in Egypt
during the summer seasons. I suppose it to
approach most in quality to the waters that are
in Europe distilled from the blossoms of Cherries,
Limes, and Acacias."

And again, in a letter to a friend he says:

" You were pleased to ask, how do the plants
subsist in Egypt without rain ? &c.

"A traveller coming to Egypt at this time, and
being unacquainted with the true reason for the
overflowing of the water, would immediately consider it as a miracle in nature. He would
imagine he beheld a sea producing vegetables.
He would behold springing from the bottom of
the sea, Sycamores, Buckthorns, Acacias, Cassias,
Willows, and Tamarisks, which form small woods,
or groves, above the surface of the water. This

is the genuine appearance of Egypt whilst it is overflown. Therefore the Egyptian plants, which consist chiefly of evergreen trees, are in no want of water."*

In Babylonia Willows were so abundant that Bochart says of the channels of the Euphrates, " The banks were so thickly lined with Willows, that Babylonia was called from them 'the Valley of Willows.'" Burckhardt also mentions a fountain in Syria, called *Ain Saffaf*, or, the Willow Fountain.

The trees on which the captives of Israel hung their harps belonged, there can be little doubt, to the species which botanists have named *Salix Babylonica*, or Weeping Willow, which grows on the banks of the Euphrates, and in other parts of Asia, and also in the North of Africa. In China it is a very favourite tree, as appears from its frequent occurrence in drawings of Chinese ornamental scenery. Throughout the same country, as well as Turkey and Algiers, it is said by Loudon to be commonly planted in cemeteries, suggesting, with its drooping branches, the idea of grief for the departed.

Gilpin considers the Weeping Willow to be a very picturesque tree. It is a perfect contrast to the Lombardy poplar. The light airy spray of the poplar rises perpendicularly ; that of the Weeping Willow is pendent : the shape of its leaf is conformable to the pensile character of the tree, and its spray, which is still lighter than that of the poplar, is more easily put into motion by a breath of air. The Weeping Willow, however, is not adapted to sublime subjects. We

* Hasselquist, " Letters from the Levant," p. 453.

WEEPING WILLOWS AT KEW.

wish it not to screen the broken buttresses and gothic windows of an abbey, nor to overshadow the battlements of a ruined castle ; these offices it resigns to the Oak, whose dignity can support them.   The Weeping Willow seeks a humble scene—some romantic footpath bridge, which it half conceals, or some glassy pool, over which it hangs its streaming foliage,

> ————and dips
> Its pendent boughs, stooping as if to drink.

In these situations it appears in character, and of course to advantage.   Nowhere is it more beautiful, than on the banks of the Thames, where are some of the finest specimens in England.

The date of the introduction of this tree into England is unknown, but it is certainly not earlier than the beginning of the last century or the close of the seventeenth, when the first tree was planted at Twickenham, either by Mr. Vernon, a merchant of Aleppo, or by Pope.   This was a favourite tree with the poet, and after his death became the object of so much curiosity that the possessor of his villa cut it down, to avoid being annoyed by persons who came to see it. Another account states that it was first planted at Kew, in 1692.

Few trees have obtained greater celebrity from their locality, than that known as Napoleon's Willow.   Loudon informs us that this tree was introduced into St. Helena from Britain, by General Beatson in 1810.   It was planted among other trees, on the side of a valley near a spring; and having attracted the notice of Napoleon, he had a seat placed under it, and used to go and sit there very frequently, and have water brought to him from the adjoining fountain. About the time of Napoleon's death in 1821, a storm shattered the Willow in pieces, and after the interment of the emperor, Madame Bertrand planted several cuttings from it on the outside of the railing which surrounded the grave. As none of these flourished, they were renewed in 1828, and from one of them, which outstripped the rest, were brought most of the cuttings which have been reared in various parts of the country. The oldest now in existence in Europe derived from this stock stands in the garden of the Roebuck Tavern on Richmond Hill, having been planted in 1823. Previously to 1810, the Willow did not grow in St. Helena; but Darwyn states that weeping Willows are now common on the banks of the rivulets, associated with so many other plants of British origin, that the imported species have excluded many of the native kinds, and given to the scenery a character decidedly British; it being only on the highest and steepest ridges that the indigenous Flora is now predominant.

So popular has the weeping Willow become, as an ornamental tree, that it is said to be commoner in almost every country than in its native habitat, the banks of the Euphrates.

THE ELDER.

SAMBUCUS NIGRA.

Natural Order—CAPRIFOLIACEÆ.

Class—PENTANDRIA.   *Order*—TRIGYNIA.

THIS tree, which possesses neither picturesque
beauty nor fragrance, comes to us recommended by
ancient authors for its numerous medicinal properties. Pliny furnishes us with a long list of the
virtues supposed to reside in the various parts of
the Elder, and our own historian of trees, Evelyn,
is no less eloquent in its praises: "If," he says,
"the medicinal properties of the leaves, bark,
berries, &c., were thoroughly known, I cannot
tell what our countrymen could ail, for which
he might not find a remedy from every hedge,
either for sickness or wound. The inner bark of
Elder applied to any burning, takes out the fire
immediately; that, or, in season, the buds boiled
in water-grewel for a breakfast, has effected won-
ders in a fever; and the decoction is admirable
to assuage inflammation. But an extract may be
composed of the berries, which is not only greatly
efficacious to assist longevity, but is a kind of
catholicon [universal remedy] against all infirmities
whatever: and of the same berries is made an
incomparable spirit, which, drunk by itself, or
mingled with wine, is not only an excellent drink,
but admirable in the dropsy. The ointment made
with the young buds and leaves in May with but-

ter, is most sovereign for aches, shrunk sinews, &c., and the flowers macerated in vinegar not only are of a grateful relish, but good to attenuate and cut raw and gross humours. And less than this could I not say (with the leave of the charitable physician), to gratify our poor woodman." Some of the above properties the Elder certainly does possess, others perhaps are imaginary; nevertheless, Elder ointment, Elder-flower tea, and Elderberry wine are still popular medicines in the country.

The Elder is a rapidly growing tree while young, and is remarkable for the stoutness of its shoots, which when a year old are as large as those of most other trees at two or three years of age. They are covered with a smooth grey bark, and contain an unusual proportion of pith, which is frequently used in electrical experiments. This pith being easily removed, young branches are often made into popguns and other toys, and on this account the Elder is sometimes called the Bore-tree. In ancient times they were made into flutes and pipes; hence the tree acquired the name Sambucus, from *sambuca*, a kind of musical instrument.* The branches do not grow so rapidly after the first year; no new pith is formed, and that which is formed already is compressed by the fresh layers of wood, so that in old stems the quantity scarcely exceeds the proportion usually found in other trees. The leaves are pinnate, slightly notched, and of a peculiarly strong and offensive odour, which is said to be unwholesome. " I do by no means,"

---

* "Countrymen believe," says Pliny, (Book xvi. Chap. xxxvii.) "that the most sonorous horns are made of Elder which has grown where it never heard the cock crow."

says Evelyn, "commend the scent of it, which is very noxious to the air; and therefore though I do not undertake that all things which sweeten

LEAF AND FLOWER OF THE ELDER.

the air are salubrious, nor all ill savours pernicious, yet, as for its beauty, so neither for its smell would I plant Elder near my habitation; since we learn from Biesius, that a certain house in Spain, seated among many Elder trees, diseased and killed almost all the inhabitants, which, when at last they were grubbed up, became a very healthy and wholesome place." Sir James Smith says, that an infusion of the leaves proves fatal to the various insects which thrive on blighted or delicate plants; nor do many of this tribe in the caterpillar state, feed on them. Cattle scarcely touch them, and the mole is driven away by their scent. Carters often place them on their horses' heads to keep off flies. The flowers are white, and grow at the extremities of the shoots, in the flat clusters which botanists call *cymes*. The berries are globular, black, and of a faint sickly taste, which no doubt often protects them from depredation. This flavour they lose when boiled and made into wine; they are said to form one of the (least injurious) ingredients of fictitious port wine. The wood of the old branches is yellow, very hard and compact, and is used for making skewers and shoemakers' pegs. The bark, which on the old branches becomes rugged, is used in Scotland as a dye. It is there called the Arn-tree.

Miss Kent observes, that the Elder is sometimes coupled with the Cypress and other trees considered to be emblematical of death or sorrow:

" The water-nymphs, that wont with her to sing and dance,
    And for her girlond olive branches bear,
  Now baleful boughs of cypress done advance :
    The muses, that were wont green bays to wear,
    Now bringen bitter Elder branches sere :
      The fatal sisters eke repent
      Her vital thread so soon was spent.
        O heavy herse !
  Mourn now, my muse, now mourn with heavy cheer :
      O careful verse ! "   SPENSER.

This notion may have originated in the tradition, that Judas Iscariot hanged himself on an
Elder-tree.

The Elder prefers a damp situation, but will
grow anywhere, bearing exposure to the sea-
breeze without receiving any injury.

" The great esteem," says Borlase, " in which the
ancient Cornish held the Elder (or Sambucus) is
very remarkable.   The Cornu-British words for it,
are *scau*, and *scauan*, and hence we have many villages,* and two ancient† families denominated. It
may at first seem to be owing to the general scarcity
of trees, that even this humble shrub was thought
considerable enough to give name to so many
places; but if we consider the great virtue of this
plant in all its several parts and stages, we shall
be convinced that few shrubs deserve a greater
regard.   It is very hardy, enduring all weathers,
suiting all soils, easily propagated by seeds and
cuttings; the medicinal use of its several parts is
extraordinary; its leaves, buds, blossoms, berries,
pith, wood, and bark, have more virtues than can
possibly have room here, without entering into
too minute detail ; the following are most obvious
and most generally applied to for relief :—The buds
and leaves, as soon as they appear, are gathered
to make baths, fomentations, and cataplasms for
wounds, and are a remedy for inflammations, &c.

---

* Boscauan-rôs, and Boscauancen, in St. Berian parish ; two
called by the name of Penscauan in St. Enodor. Enyscauan in St.
Denis ; Lescauan in Sheviock ; Fentonscauan, name of a water, in
St. Ives ; Trescau, formerly the most considerable village in the
Scilly Isles ; Trescau in Brêg, &c.

† That of the Right Honourable Lord Viscount Falmouth, called
Boscawen, and the Scawens of Molenik in St. Germans, and of
Carshalton in Surrey.

As soon as the flower-buds come on, they serve
to make a pickle of very good flavour; the flowers
at their opening, infused, communicate their taste
and smell to vinegar,—infused and let to stand
in best Florence oil, excellent to be laid over
bruises, and external swellings; and taken inter-
nally, very healing and cooling; the flowers in
their natural state, are very sudorific, and assuage
pains; distilled with simple water, make a sweet
cooling wash for the face in summer, which takes
off inflammations of the eyes (as a collyrium), is
good for the wind in children, and a very innocent vehicle in fevers; distilled in spirits, it as-
suages cholical pains in adult persons; and there
is a spirit to be drawn from the Elder, which the
late Duke of Somerset, who married the heiress
of Piercy, took for the gout, as I am informed,
with success. When the berries are ripe, they
make a very wholesome syrup in colds and fevers;
and some make wines of them, by mixing Rhenish
or other white wines. Of the younger sappy
branches the bark, pared off close to the wood,
makes a salve efficacious beyond most others for
scalds. This inner bark is also very salutary in
dropsies, says Mr. Ray. The wood is close-
grained, sweet, and cleanly, and beyond any
other chosen by butchers for skewers, as least
affecting their flesh; it is very beautiful also
for turners'-ware, and fineering, and for toys of as
neat a polish as box; and the very pith of this
useful shrub is proper to cool, and make ulcers
and wounds digest. More uses than these may
occur by way of medicine, but the above are
perhaps more than sufficient to shew that the
Cornu-Britons did not denominate places and

persons from this seemingly contemptible shrub
without great propriety; its peculiar properties
are not to be wondered at, though numerous;
they are indeed chiefly medicinal, and those of
other plants are sometimes principally nutritious and domestic. Nature has differently distributed her bounties among plants, and placed
them together sometimes in great numbers.
The Palm-tree, as Strabo says, has three hundred and sixty uses, and the Cocoa or Coker-
nut-tree yields wine, bread, milk, oil, sugar, salt,
vinegar, tinctures, tans, spices, thread, needles,
linen-cloth, cups, dishes, baskets, mats, um-
brellas, paper, brooms, ropes, sails, and almost all
that belong to the rigging of a ship, if we may
believe Fr. Hernandez, and other authors. Beside this *Sambucus aquatilis seu palustris*, we
have another sort, which we call Scau-au-Cûz, or
the Elder of the Wood;—some call it Maiden
Elder. Its uses have not been hitherto discovered to be as various and salutary as those of
the foregoing, but its wood is more flexible, and
will divide lengthways, as perfectly almost as
whalebone, and is therefore much coveted by
joyners."

THE WOODBINE, or HONEYSUCKLE.

Lonicera periclymenum.

Natural Order—Caprifoliaceæ.

Class—Pentandria.   *Order*—Monogynia.

No British shrub claims our favourable notice so early in the season as the Honeysuckle;
for even before the earliest Snowdrop has ventured to pierce the unthawed earth, we may discover in the sheltered wood or hedge-bank its
wiry stems, throwing out at every joint tufts of
tender green foliage. In this state it is even
richer in promise than the fully-expanded winter
flowers, for belonging as it does to the brightest
days of summer, its opening buds carry us away
at once to the genial season when the fields are
decked with their gayest attire, and the air
loaded with the most delicious perfumes, among
which its own fragrance is to occupy no mean
position. Later in the year it engages our attention by its twisting stems clinging for support
to some lustier neighbour, and climbing with
undeviating accuracy from left to right until it has
overtopped its friendly support, when it asserts
its independence, loses a good deal of its twining
character, and displays its numerous clusters of
trumpet-shaped flowers.

As its coil of stem, when once formed, never
materially enlarges, and is too tough to yield to
the expanding force of the tree around which it

twines, it is a mischievous neighbour to the young
sapling, stopping its growth, and forming a spiral
channel in its bark, which is eventually the source
of disease and death.   Cowper's description of the
tree is therefore more accurate than that of Shakspeare : —

> " As Woodbine weds the plant within her reach,
> Rough Elm, or smooth-grain'd Ash, or glossy Beech,
> In spiral rings ascends the trunk, and lays
> Her golden tassels on the leafy sprays ;
> But does a mischief while she lends a grace,
> Straitening its growth by such a strict embrace."
>
> COWPER.

> " So doth the Woodbine—the sweet Honeysuckle,
> *Gently* entwist the Maple."
>
> SHAKSPEARE.

The Woodbine is scarcely less a favourite with
the poets than the Hawthorn.   Milton eulogizes
it under a name which belongs of right to the
Wild Briar, namely, Eglantine :—

> " Through the sweet-briar, or the vine,
> Or the twisted Eglantine."

Wordsworth's description is strictly true to
nature :—

> " So, pleased with purple clusters to entwine
> Some lofty Elm-tree, mounts the daring Vine ;
> The Woodbine so, with spiral grace, and breathes
> Wide-spreading odours from its flowery wreaths."

And again :

> " Brought from the woods the Honeysuckle twines
> Around the porch, and seems in that trim place
> A plant no longer wild."

The Honeysuckle is in most luxuriant bloom in

June; its flowers, copiously stored with honey, are then rifled by such insects as are furnished with a long proboscis; while others which cannot reach to the bottom of its curved tubes, obtain their booty by piercing the base, a method which is successfully pursued with other tubular flowers, such as the Jasmine. To the flowers succeed bunches of scarlet berries, which are clammy to the touch, glutinous, and sweet to the taste, but mawkish. In October the Woodbine, with praiseworthy perseverance, endeavours to impart a grace to the fading year by producing a new crop of flowers, which, though not so luxuriant nor so numerous as the first, are quite as fragrant. Clusters of flowers and of ripe berries may then be found on the same twig, uniting autumn with summer as the early foliage united winter with spring.

The name Lonicera was given to it in honour of Lonicer, a German: Periclymenum is a Greek compound, and signifies winding about: Woodbine is evidently a corruption of Woodbind, and Honeysuckle has reference to the custom among children of sucking honey out of the flowers.

The Honeysuckle is propagated either by cuttings or layers; but a yet readier way to secure a stock of the common variety is to collect young rooted plants in the woods and hedges, taking care to select the month of October or November for the operation; for, if transplanted at this season, they rarely fail to grow.

Many foreign species of Honeysuckle are cultivated, but these belong to the garden, rather than to the woodland. One species, *Lonicera*

T

Caprifolium, Perfoliate Honeysuckle, is supposed
by some to be a native of Britain; it may be
distinguished by having its pairs of opposite
leaves united at their bases, and forming a kind of
cup, through which the stem passes.

WAYFARING-TREE.

Viburnum Lantana.

GUELDER ROSE.

Viburnum Opulus.

Natural Order—Caprifoliaceæ.

Class—Pentandria.   *Order*—Monogynia.

These two shrubs, and the common garden Laurustinus, *Viburnum Tinus*, agree in having a funnel-shaped corolla of one petal, and a calyx divided into five deep segments, which remains attached to the fruit, a single-seeded berry, until the latter is ripe.

The Wayfaring-tree may easily be distinguished at all seasons by its numerous pliant mealy branches, which in winter are ornamented by hoary button-like buds, and in summer are clothed by heart-shaped leaves, covered with mealy down. The flowers are white, and grow in clusters at the extremities of the shoots, and are succeeded by flattened berries, which as they ripen become red, and finally black. A modern poet, William Howitt, captivated by the pleasing name, has addressed an Ode to the Wayfaring-tree, and eulogises its "coronets of fragrant snow,"—a metaphor the propriety of which any one who knows the tree will find it difficult to discover, the flowers being by no means attractive. It would seem to owe its name to the soiled appearance of its leaves, which, wherever the tree is growing, give one the notion of its having been powdered with dust from

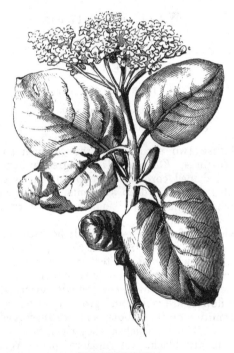

WAYFARING-TREE.

the highway. The season when this tree is most conspicuous is when the berries are partially ripe; for then the scarlet and black berries growing in the same clusters present a very singular appearance. Loudon tells us that in Germany the young shoots are employed in basket-making, and for tying faggots and other packages; and those of two or three years old are used for the stems of tobacco-pipes. The berries are used in Switzerland for making ink.

The Guelder Rose is a much prettier shrub than the preceding. In its wild state it never attains the dimensions of a tree, but is nevertheless a highly ornamental hedge-plant. The tree is smooth in every part, the leaves are large, three-lobed, and serrated. The flowers are of a brilliant white, and much more conspicuous than those of the Way-faring-tree, growing in clusters which have the outer flowers destitute of stamens and pistils, but furnished with large and showy petals. In August and September its bunches of coral berries are very ornamental, and in October it is yet more conspicuous. The foliage then assumes a deep crimson purple hue, and if the tree, as is frequently the case, be associated with the bright yellow foliage of the Maple, no garden, however richly stocked with the most showy flowers, presents so gorgeous an appearance. It is said that in Siberia the berries are made into paste with honey and flower, and eaten as food, but this is scarcely credible, so exceedingly offensive is the odour which they emit. Long after the trees have been stripped of every leaf, the clusters of crimson berries at the extremities of the branches retain their bright appearance, drooping indeed, and shrivelled with the frost, but nevertheless very attractive to the eye. The wood, like that of the Spindle-tree and Cornel, is much used for making skewers; indeed the name of Dog-timber, which properly belongs to the Cornel, is in may places given to all three of these shrubs.

GUELDER ROSE.

The Guelder Rose-tree of gardens,

> "Tall,
> And throwing up into the darkest gloom
> Of neighbouring Cypress, or more sable Yew,
> Her silver globes, light as the foamy surf
> That the wind severs from the broken wave,"

is a variety of this species, differing from the usual character in having all its flowers barren and

crowded together in the form of a globe; hence
it derives its name of Snowball-tree. This is a
very desirable tree to plant among other shrubs,
both for the sake of its flowers, and on account of
the rich purple hue of its foliage in autumn.

THE IVY.

HEDERA HELIX.

Natural Order—ARALIACEÆ.

Class—PENTANDRIA.   *Order*—MONOGYNIA.

THE Ivy is a tree of very ancient repute, occupying a prominent place in the Mythology of the Greeks and Romans, and applied to purposes which were deemed the most honourable. The warrior-god Bacchus had his brows and spear decked with Ivy: the people of Thrace adorned their armour with the foliage of the same tree, and an Ivy crown was the highest prize that was awarded to a successful poet. The Grecian priests presented newly-married couples with a wreath of Ivy, as a symbol of the closeness of the tie which ought to bind them together; and it continues a favourite emblem of constancy among the moderns. Owing to a property, which it is supposed to possess, of absorbing nourishment, by means of its root-like tendrils, from the trees to which it clings, some consider its friendship not strictly disinterested:

> " He was
> The Ivy which had hid my princely trunk,
> And suck'd my verdure out."

With many, the Ivy is the tree peculiarly dedicated to gloom; its foliage is heavy, and of a som-

bre hue; it shews its flowers and strives to be as
gay as it can, when almost every other tree has
finished its summer course; it loves to creep over
sepulchres and ruined buildings, as even Pliny has
remarked; it courts retirement and the shade, and
if it does sometimes grow on a sunny bank, it
seems sickly and ill at ease, rarely rising from the
ground unless it can avail itself of the support
afforded by some decaying tree that has little
foliage of its own. But I am by no means disposed to allow that the Ivy deserves this un-
amiable character; for though the facts are true
enough, a very different inference may be drawn
from them. It certainly does grow most luxuriantly over the ruined walls of buildings, but, with
its verdure "never sere," rather takes from their
gloominess than adds to it; and if it does begin
its chilly summer when winter reigns over all
the forest beside, surely it deserves not a little
gratitude for exerting itself to prolong the season of flowers, and to spin out the existence
of the myriads of insects which would certainly
perish were it not for the copious supply of
honey afforded by its abundant clusters of
flowers. Even if the accusation be true, that
it is never at ease unless it be getting up in
the world, its ambition is scarcely to be blamed,
for it mostly avails itself of the support afforded by trees whose own vigour is irrecover-
ably gone, and which, but for the borrowed
verdure of the visitor, would be stark and unsightly trunks.

As an ingredient in the landscape it does not
need any apologist. The opinion of Gilpin, the
greatest authority in such matters, is impartial and

decisive: "Ivy is another mischief incident to trees, which has a good effect. It gives great richness to an old trunk, both by its stem, which winds round it in thick, hairy, irregular volumes; and by its leaf, which either decks the furrowed bark, or creeps among the branches, or hangs carelessly from them. In all these circumstances it unites with the mosses and other furniture of the tree in adorning and enriching it; but when it gathers into a heavy body, which is often the case, it becomes rather a deformity. In autumn I have seen a beautiful contrast between a bush of Ivy, which had completely invested a pollard Oak, and the dark brown tint of the withered leaves, which still held possession of the branches. In the spring also we sometimes have a pleasing appearance of a similar kind. About the end of April, when the foliage of the Oak is just beginning to expand, its varied tints are often delightfully contrasted with the deep green of an Ivy-bush, which has overspread the body and larger limbs of the tree; and the contrast has been still more beautiful, when the limbs are covered, as we sometimes see them, with tufts of brimstone-coloured moss [lichen]."

The amiable author of the "Journal of a Naturalist," has the following pleasing remarks on Ivy growing in another situation:—" As a lover of the lone Ivy-mantled ruin," he says, " I have often questioned with myself the cause and basis of my regard for that which was but a fragment of what might have been formerly splendid, and intrinsically possessed but little to engage admiration, yet wreathed in the verdure of the Ivy, was admired; but was never satisfied, perhaps

unwilling to admit the answer that my mind
seemed to give. The Ivy is a dependent plant,
and delights in waste and ruin. We do not often
tolerate its growth when the building is in repair
and perfect, but, if time dilapidate the edifice,
the Ivy takes possession of the fragment, and we
call it beautiful; it adorns the castle, but is an
indispensable requisite to the remains of the monastic pile. There is an abbey in the north of
England which has been venerated by all its late
possessors. It is trimmed, made neat, and looks,
perhaps, much as it did formerly, except being in
ruins. The situation is exquisite, the remains are
splendid, yet with many it fails to excite such interest as it should do. It is a bare reality. A
ruin in the west of England once interested me
greatly. The design of revisiting and drawing it
was expressed at the time. A few days only
elapsed; but the inhabitant of a neighbouring
cottage had most kindly laboured hard in the
interval, and pulled down 'all the nasty Ivy, that
the gentleman might see the ruin.' He did see it,
but every charm had departed. These two instances, from many that might be advanced, manifest that Ivy most frequently gives to these
ancient edifices the idea of beauty, and contributes
chiefly to influence our feelings when viewing
them. The ruins of a fortress or warlike tower
may often historically interest us, from the renown
of its founder or its possessor, some scene trans-
acted, some villain punished, hero triumphant, or
cause promoted to which we wished success; but
the quiet, secluded monastic cell or chapel has
no tale to tell; history hardly stays to note even
its founder's name; and all the rest is doubt and

"HOLY-WELL" AT TRELILL, NEAR HELSTON, CORNWALL.

darkness; yet, shrouded in its Ivy-folds, we reverence the remains, we call it picturesque, we
draw, we engrave, we lithograph the ruin.   We do

not regard this Ivy as a relic of ancient days; as
having shadowed the religious recluse, and with
it often, doubtless, piety and faith; for it did not
hang around the building in old time, but is
comparatively a modern upstart, a sharer of monastic spoils, a usurper of that which has been
abandoned by another. The tendril, pendent
from the orient window, lightly defined in the
ray which it excludes, twining with graceful ease
round some slender shaft, or woven amid the
tracery of the florid arch, is elegantly ornamental, and gives embellishment to beauty; but the
main body of the Ivy is dark, sombre, massy;
yet, strip it from the pile, and we call it sacri-
lege, the interest of the whole is at an end, the
effect ceases,

'One moment seen, then lost for ever.'

Yet what did the Ivy effect? what has departed
with it? This evanescent charm, perhaps, consists in the obscurity, in the sobriety of light it
occasioned, in hiding the bare reality, and giving
to fancy and imagination room to expand, a play-
thing to amuse them."

Ivy is often associated with Holly and other
evergreens in the decoration of our churches
at Christmas, but for no other reason that I am
aware of than that it retains its freshness for
a considerable time, and that its dark berries
contrast well with the bright scarlet berries of
the Holly.

The Ivy is confined to temperate climates, but
grows wild neither in America nor Australia.
About Smyrna in Asia Minor it is very common,
forming the greatest part of the hedges, and

ornamenting every garden.

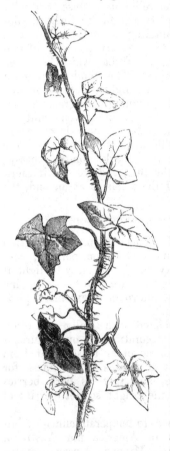

In the Himalayas it is also very abundant, producing yellow berries. This variety is supposed to be the plant which was held in such high repute among the Greeks and Romans.

No British plant varies so greatly in its habit at different periods of growth and in the shape of its leaves as the Ivy. In its infancy it is a brittle, climbing stem, furnished with alternate three or five-lobed leaves, which are light green, or of a reddish tinge, with white ribs. As it increases in size the lobes of the leaves become wider, and the stem loses its brittle character. If it can find no support, it creeps along the ground, sending into the earth, from the lower part of the stem, tufts of fibrous roots. These are to be distinguished from the tufted fibres, by the aid of which the plant clings to a wall or trunk of a tree. The former are proper roots, and are only pushed forth from those parts of the stem which are opposite to the leaves, and only appear when they can be employed with advantage in the support of the plant. The tufts of fibres, on the contrary, are produced from all parts of the stem which are nearest to the wall or tree, and are invariably absent when the plant crawls along the ground and therefore has no use for them. Consequently the opinion that Ivy is injurious to trees, by inserting its roots into the bark and absorbing their juices, is erroneous. It may sometimes happen that a tree is clasped too closely by matted Ivy-bands, and is thus prevented from developing its full growth, or, more rarely still, the weight of its massy head may overstrain the branch which supports it, and be the occasion of ruin to both; but, except under such circumstances, it does no mischief.

A remarkable instance of the compressing power of Ivy is cited in the "Gardeners' Chronicle," proving that a netted mass of Ivy does not simply prevent the expansion of the body which it clasps, but as its stems increase in bulk, actually contracts the space enclosed. On removing some Ivy from an old house at Carshalton, it was discovered that a thick leaden water-pipe had been in many places deeply indented and in some places squeezed flat by the stem of the plant. Trees that have long been coated with a net-work of Ivy should not be stripped all at once, lest they should be injured by sudden exposure to cold; and when it is desired to keep young trees in plantations

clear of Ivy, the best plan is, not to cut through
the stems of the intruder, as generally practised, but to detach them as carefully as possible from the trees, and to let them fall back.
They will thus lie on the ground, and continue to
grow in the same direction in which they were
laid. Otherwise, new shoots will spring up from
the roots which have been deprived of their
leading stems, and it will soon be necessary to
repeat the process. When Ivy grows over buildings its effects depend on the nature of the structure : if the masonry be solid no mischief can
ensue, as the climbing shoots will bind and
strengthen, without attempting to penetrate : but
if the structure be loose and crumbling, or if
earth be lodged here and there, it is very likely
that roots will be formed wherever they find
a convenient soil, and, as they increase in size,
will penetrate into the mass, and dislodge the
constituent parts. A striking example of the
pernicious effects of Ivy on a structure of this
kind occurred some years since in a remote county.
At a period of great agricultural distress, a gentleman, in order to furnish the poor with employment, resolved to enclose his park, which was
seven miles in circumference, with a stone wall.
The mason who undertook the contract happened to be an unprincipled man, and instead of
fulfilling his engagement of building a solid wall
of stone, erected it with a double facing of the
material named, and filled the interstice with earth
and rubbish. When completed, it appeared to be
an honest stone wall, but in a few years Ivy
climbed to the top in many places, sent down its
roots into the earth, and these, as they enlarged,

thrust out the stones which constituted the facings, and revealed the iniquity of the contractor. The present proprietor is subjected to a heavy annual expense in repairing the mischief done by a plant, which, if the structure had been what it appeared to be, would have added greatly to its strength and durability.

It has long been a disputed question, whether Ivy growing against the side of a house renders it damp or otherwise. Dr. Lindley thus pronounces his opinion, formed from a comparison of various conflicting statements made in the " Gardeners' Chronicle:" " Ivy may render a house damp by retaining snow in winter, which changes to water, trickles down the walls, and never thoroughly evaporates. But this is of rare occurrence, and may be prevented by beating the Ivy after snow-storms, and will only be an inconvenience when houses are built with mud. No doubt, when walls are not of sound brickwork, or of some other hard materials, the Ivy may introduce its roots into the masonry, and thus do mischief, allowing water to run down its branches and to follow them into the crevices where they have insinuated themselves. But in all cases of well-built houses, we are convinced that Ivy is beneficial, so far as keeping the walls dry."

When Ivy has mounted to the summit of its support, its character and habit undergo a material alteration; it is no longer a climbing stem with lobed leaves, but sends out erect branches of tufted foliage and becomes a round-headed bush. Neither roots nor tendrils are formed on the stems; and the dark, glossy leaves preserve an even edge, unbroken by any indentation, but

still varying considerably in width.   The height
at which this alteration takes place varies from

BRANCH OF IVY.

a few feet to a hundred, for it seems to require
not so much an elevated tract of atmosphere, as
free access to light.   Its upward growth now
rarely exceeds a few feet, but it produces abundance of leaves and flowers.   The latter are formed
in terminal heads; each flower is furnished with
a separate stalk, and comprises five green petals,
five stamens, and one pistil.

When the month of October happens to be

enlivened by but few bright days, an Ivy-bush
in full bloom suggests the idea of anything
but gloom. All the trees of the forest are
plainly intimating that their glory is in the
wane; a few pale flowers are scattered here and
there, evidently the produce of exhausted plants
—the whole insect world, with the exception of
drony evening beetles, has either perished or
retired to secure winter quarters,—when, after
some days of storm and cloud, there comes
a flash of calm, clear sunshine: then, hasten
to the nearest Ivy-bush, and be convinced that
summer has not taken its departure without
giving one day as an earnest that it will come
again. Every twig of Ivy terminates in a cluster of fresh, timely flowers, which, sober though
they may be in hue, shew no symptom of decay, and, at the same time, lengthen the existence of myriads of insects. The Red Admiral
butterfly especially, is sure to be a guest at this
banquet, but is far from being alone; the Painted
Lady regales herself close by; and flies of all
sizes and shapes, hornets, wasps, bees, all flock
hither in wonderful harmony to enjoy once more
a full feast before they submit to the necessity
of their long winter's fast.

A few months later, and the banquet is spread
again on the same table for another winged tribe.
Blackbirds, thrushes, and wood-pigeons know well
where Ivy-berries grow, and, now that they have
stripped the Hawthorn and Mistletoe bare, resort
to the Ivy-bush in quest of food by day, and
shelter by night, and many a cluster of barren
stems shews how keen was their appetite; while
the abundance yet left tells as plainly of the

ample provision that their Heavenly Father had
made for them, during even the most inclement

IVY-BERRIES.

period of the year.   It is a fact well worthy of
note, that Ivy-berries are never injured by frost,
however severe the winter may be.

Although the Ivy never bears flowers or assumes a bushy habit until it has had an opportunity of indulging its climbing propensities, yet,
by proper management, it may be made highly
ornamental as a standard shrub. For this purpose, plants that have mounted to the top of a
hedge-bank, and have there rooted, should be
taken up in winter, and carefully removed to their
new destination, when, though they may perhaps
throw off all their leaves (a tolerably sure sign

of healthy action in any transplanted tree), they
will soon recover; for it appears that the bushy
branches, when once formed, never revert to the
habit of the young plant.

The principal use of Ivy is that already
mentioned, namely, of covering the walls of buildings. Planted against the side of a house,
where there are no windows, it is not only
ornamental, but keeps out heat in summer,
and cold in winter; but when it climbs round
windows, it is likely to be the means of intro-
ducing earwigs and other insects into the house.
The variety called Irish Ivy, which has large
leaves, and grows rapidly, is the best adapted
for covering masonry.

The leaves and tender branches are eaten by
sheep and deer in times of scarcity. The wood
is soft and porous, and when cut into thin slices
is used in filtering liquids. The roots are employed by leather-cutters to sharpen their knives
on. A fragrant resin exudes from the old stems
if wounded, which, Walton says, makes bait attractive to fish. A substance called hederine
may be extracted from this, which in India is
used as a medicine.

The largest plants of Ivy recorded by Loudon
as growing in England are at Brockley Hall in
Somersetshire, attached to old trees; one of these
has a stem nearly eleven inches in diameter, another, nearly twelve inches; another, yet more
remarkable one, at Morpeth, grows out of a crevice
in a rough stone wall by a cottage, which at the
height of nine feet from the ground is one foot
seven inches and a half in girth. De Candolle
speaks of another at Gigean near Montpelier,

which was six feet in circumference at the base,
and divided into two great trunks, from two to
five feet in girth; these trunks grew at first erect
and afterwards rested upon a wall. The branches
of this plant covered seventy-two square yards,
and the whole height was eighteen feet when De
Candolle examined it, but it had been larger; he
estimated its age at four hundred and thirty-three
years.

Near the ruins of Fountain's Abbey is an
aged Ash-tree, which is encircled by beautiful
Ivy in rich luxuriance, the stem of the latter
having attained the size of three feet two inches
in girth.

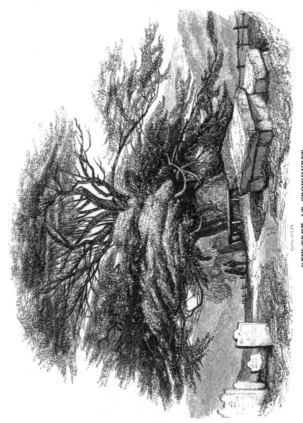

YEW-TREE AT CROWHURST.

THE YEW.

TAXUS BACCATA.

Natural Order—TAXACEÆ.

Class—DIŒCIA.   *Order*—MONADELPHIA.

THE Yew-tree, "neither verdant, nor graceful,
but gloomy, terrible, and sapless," to judge from
Pliny's description, is a tree of evil omen. Not
only were the berries deemed poisonous, but
vessels made of the wood were said to impart the
same property to wine kept in them, and it was
considered more than hazardous to sleep or take
food under the shade of its branches. The very
name for the poison with which arrows were
armed, *toxica,* was, according to the same author,
a corruption of *taxica,* from *taxus,* the Latin name
of the tree. Virgil agrees with Pliny in con-
demning the Yew; he calls it a noxious tree, and
recommends that it should not be allowed to
stand near bee-hives. Other authors, ancient and
modern, join in assigning to it properties deadly
to various kinds of animals. No wonder then
that the frequent appearance of the Yew in
churchyards should have suggested the idea that
it was planted in such situations as an emblem
of death, and a fit shelter for the dead. That
the Yew was commonly planted by our forefathers in churchyards, there can be no doubt,
for there are yet in existence a vast number of
these trees so planted many centuries since; but

there is far greater probability that at the period
when crosses were erected in these sacred spots as
emblems of the victory over death achieved by
the Author of our faith, the Yew-tree was sta-
tioned not far off, to symbolize, by its durability
and slowly altering features, the patient waiting
for the resurrection, by those who committed the
bodies of their friends to the ground in hope.
Heathens indeed might with propriety have selected the most deadly of trees to represent the
character of what they might well consider a merciless destroyer ; but such a feeling could have
had no place with sober Christians. They, on
the other hand, would regard the perpetual verdure which overshadowed the remains of their
forefathers, and was shortly destined to canopy
their own, as the most fitting expression of their
faith in the immortality of the soul. Generation
after generation might be gathered to their fathers,
the Yew-tree proclaiming to those who remained,
that all, like the ever-green, unchanging Yew,
were yet living, in another world, the life which
had been the object of their desire. The Yew,
then, we may safely conclude, is not an unmean-
ing decoration of our churchyards, much less a
heathenish symbol, or, as some will have it, a tree
planted with superstitious feelings, but an appropriate religious emblem :—

> " Of vast circumference, and gloom profound,
>   This solitary Tree ! A living thing
>   Produced too slowly ever to decay ;
>   Of form and aspect too magnificent
>   To be destroyed."
>   WORDSWORTH.

Miss Kent quotes from Dr. Hunter a passage

which quite supports this view. " Dr. Hunter
thinks the best reason to be given for planting
the Yew in churchyards is, that the branches were
often carried in procession on Palm Sunday,
instead of Palm." It is still customary in Ireland
for the peasants to wear sprigs of Yew in their
hats from that day until Easter-day. " Our forefathers," says Martyn, " were particularly careful
to preserve this funereal tree, whose branches it
was usual to carry in solemn procession to the
grave, and afterwards to deposit therein under the
bodies of their departed friends. Our learned
Ray says, that our ancestors planted the Yew in
churchyards, because it was an evergreen tree, as
a symbol of that immortality which they hoped
and expected for the persons there deposited.
For the same reason, this and other evergreen
trees are even yet carried in funerals, and thrown
into the grave with the body; in some parts of
England and in Wales, they are planted with
flowers upon the grave itself."* Shakspeare
alludes to a similar custom :—

> " My shroud of white, stuck all with Yew,
>   O prepare it ! "

Phillips quotes a table taken from the ancient
laws of Wales, from which it appears, that some
trees were solemnly dedicated to religious purposes, and were consequently more highly valued
than others :—

> " A consecrated Yew, its value is a pound ;
>   An Oak, its value is six score pence.
>   *   *   *   *
>   Fifteen pence is the value of a wood Yew-tree."

---

* This custom is still observed in the north of Devon.

Dr. Aikin thinks it probable, that the Yew was planted in churchyards for the sake of furnishing boughs to decorate the church at Christmas, but Miss Kent has shewn, by a quotation from Brand's "Popular Antiquities," that the Yew was rarely used except in default of other evergreens: " Had a tree," she says, " been planted in churchyards for that use, it would more probably have been the Holly, which was never omitted." Herrick speaks of the Yew as expressly appropriate to the season of Easter:—

> " The Holly hitherto did sway,
> Let Box now domineer,
> Until the dancing Easter-day,
> Or Easter's-eve appear.

> " Then youthful Box, which now hath grace
> Your houses to renew,
> Grown old, surrender must his place
> Unto the crisped Yew."

Perhaps the favourite opinion is, that Yews were planted in such situations to afford a supply of wood for making bows. The long-bow, it is well known, was at the period of the battles of Cressy, Poictiers and Agincourt, the national weapon of England. Statutes were passed by many of our sovereigns forbidding the exportation of Yew wood, and obliging Venetian and other merchant ships to import ten bow-staves with every butt of wine, and by an act passed in the reign of Edward IV., every Englishman residing in Ireland was expressly ordered to have an English bow of his own height, made of Yew, or some other wood. The best bows, however, were not made of native wood, foreign Yew being thought so much superior, that a bow of it sold for six

shillings and eight-pence, when the bow of English
wood cost only two shillings. It does not appear,
therefore, that the English Yew-tree was sufficiently prized for its wood, to need the protection
of a churchyard; and if it had been highly valued,
we should rather expect to find traces of extensive
plantations, than solitary trees in churchyards,
which, after all, were very inappropriate places to
plant trees intended to be applied to warlike
purposes.

Mr. Bowman has written an article in the
" Magazine of Natural History," in which he
states it as his opinion, that the Ancient Britons,
before the introduction of Christianity, planted
Yew-trees near their temples from the same
superstitious motives that actuated the Canaan-
ites, who, we are told, were in the habit of performing their idolatrous rites in groves. When
Augustine was sent by Gregory the Great to
preach Christianity in Britain, he was particularly
enjoined not to destroy the heathen temples, but
only to remove the images, to wash the walls with
holy water, to erect altars, &c., and so convert
them into Christian churches. The Yew-trees
consequently were allowed to remain, as not
necessarily conveying any erroneous impression.
There are still in existence Yews which, in all
probability, were venerable trees before the introduction of Christianity.

Mr. Bree, too, is of opinion, that churches were
frequently built in Yew-groves or near old Yew-
trees rather than that the trees were planted in the
churchyards after the churches were built. Such,
probably, was often the case; but whether the
church or the tree were the first to occupy the

site, our Christian forefathers cannot with propriety be said to have sanctioned the custom
either from superstitious feelings or for the sake
of supplying the demand for bow-staves.

The Yew is a native of most of the temperate
parts of Europe and Asia, growing in its wild
state in situations little exposed to the direct rays
of the sun, such as the north side of steep hills,
or among tall trees, and, according to Loudon,
always in a clayey, loamy, or calcareous soil,
which is naturally moist. The same author also
states that the Yew is rather a solitary than a
social tree, being generally found either alone or
with trees of a different species. This is, however,
far from being always the case, for the Yew-tree
Island in Loch Lomond, some twenty years ago,
furnished three hundred Yews for the axe; and
there are still a number of fine specimens on it;
it is also abundant on the north side of the mountains in the same neighbourhood. There are also a
great number of these trees on the cliffs near Coomb
Martin in the north of Devon, growing in places
which are accessible only to birds. But the most
remarkable assemblage of Yews in Great Britain,
is at a place called Kingly Bottom, about four
miles from Chichester. As to when, or by whom
they were planted, or indeed whether they were
planted by the hand of man at all, history is silent.
They are about two hundred in number; one half
of them form a dense, dark grove, in the depth of
the bottom; the remainder, smaller ones, are
scattered over the sides of the valley, intermingled
with fine plants of Juniper and Holly. The trunks
of the largest vary from twelve to twenty feet in
circumference, at three feet from the ground;

their greatest height is about forty feet, and their extreme spread sixty feet in diameter. Tradition fixes their age at nine hundred years. About two years since a lady happened to be driving through the grove, and observed something like fire at the base of one of the trunks, which somewhat surprised her, as no one is permitted to light a fire under the trees, and moreover, there was not a living creature to be seen within the confines of the valley. She approached the spot, and found that the trunk of one of the largest trees was a glowing mass, whilst the branches were as fresh and the leaves as green, as those of the uninjured trees around. She drove back to the village (about a mile distant) and informed the person who had charge of the trees, of what she had observed. He immediately sent a man towards the spot, but he was so awe-stricken by the solitude of the vale and the unaccountable fire, that he was afraid to approach it. Others, however, soon came with a good supply of water, and after many hours extinguished the fire; but the tree fell. My informant saw it a day or two afterwards, when it presented a very extraordinary appearance. Although the entire trunk from the commencement of the branches to beneath the surface of the ground was converted into charcoal, the crown of the tree seemed entirely unaffected by the fire. The cause of the fire was not ascertained. It could not have been occasioned by friction: but as one side of the tree was decayed, it is probable that fire was actually applied to it. Spontaneous combustion may, however, have taken place amidst the rotten wood and other materials collected in the hollow.

There are several earthworks supposed to be
British on the surrounding hills; and on the
highest ridge, four or five hundred feet above the
bottom, are some tumuli or cairns, which are
supposed to be the tombs of "Danish Sea-kings."
The old name of the place is said to be "Kings'-
slain Bottom."* The vicinity of these ancient
Yews to the British remains would seem to favour
Mr. Bowman's view mentioned above.

The Yew-tree is characterised by a trunk peculiarly suggestive of massiveness and solidity,
not being covered, like the trunks of most other
trees, with a splitting bark, but seemingly composed of a number of smooth stems fused together.
The bark itself is of a reddish brown hue, and
scales off in thin plates. At the height of a few
feet from the ground, it sends out numerous
horizontal branches, which spread in all directions, and are densely clothed with tough twigs,
which are leafy throughout their whole extent, or
nearly so. The leaves are thickly set on two opposite sides of the stem, very narrow, slightly recurved, dark green, and shining above, but paler
below. The young shoots of the Yew are subject
to a disease, the effect of which is a conical bunch
of succulent leaves at the extremity of the twigs;
this, when it has grown to about an inch in length,
withers and dies off. The flowers, which are of
two kinds, and grow on separate trees, appear
among the leaves, and on the under side of the
twigs. The barren flowers are the most numerous,
appearing in the form of membranous scaly buds,
from the centre of each of which protrudes a slen-

* I am indebted for the above account of the Kingly Bottom
Yews to Dr. N. Tyacke.

der column, terminating in a tuft of stamens. The fertile flower resembles a minute acorn, the cup of which swells, and when ripe has the appearance of red cornelian, enclosing an oval brown

LEAF AND FLOWER OF THE YEW.

nut, the summit of which is uncovered. These berries, if berries they may be called, droop when ripe, and contain a sweet glutinous juice. They are of a mawkish, disagreeable taste, but are eaten with impunity by children, and greedily devoured

II.   X

by wasps and moths. The nut contains a kernel, which is eatable, and has an agreeable flavour like that of the Stone Pine. The leaves are poisonous, though to what extent is a disputed question; but of this there can be no doubt, that their effects on the human frame are deadly, and that to give them to cattle is a perilous experiment. Instances are on record of cattle eating them with impunity, mixed with other fodder: but whether in a green or half-dry state, they are highly dangerous. It appears from all accounts that the poison is more virulent in the young shoots than in any other part of the tree, but that it exists in greater or less quantities both in the fully expanded leaves and in the green bark.

" As to its picturesque perfections," says Gilpin, " I profess myself (contrary I suppose to general opinion) a great admirer of its form and foliage. The Yew is of all trees the most tonsile. Hence all the indignities it suffers. We everywhere see it cut and metamorphosed into such a variety of deformities, that we are hardly brought to conceive it has a natural shape, or the power which other trees have of hanging carelessly and negligently. Yet it has this power in a very eminent degree; and in a state of nature, except in exposed situations, is perhaps one of the most beautiful evergreens we have. Indeed, I know not whether, all things considered, it is not superior to the Cedar of Lebanon itself, I mean to such meagre representations of that noble plant as we have in England. The same soil which cramps the Cedar is congenial to the Yew. It is but seldom, however, that we see the Yew in perfection. In New Forest it formerly abounded,

but it is now much scarcer. It does not rank
among timber-trees; and being thus in a degree
unprivileged, and unprotected by forest laws, it
has often been made booty of by those who durst
not lay violent hands on the Oak or the Ash.
But still in many parts of the forest some noble
specimens of this tree are left. One I have
often visited, which is a tree of peculiar beauty.
It immediately divides into several massy limbs,
each of which, hanging in grand, loose foliage,
spreads over a large compass of ground, and yet
the whole tree forms a close, compact body; that
is, its boughs are not so separated as to break into
distinct parts. It cannot boast the size of the
Yew at Fotheringal, near Taymouth, in Scotland, which measures fifty-six feet and a half in
circumference; nor indeed the size of many others
on record; but it has sufficient size for all the
purposes of landscape, and in point of picturesque
beauty it probably equals any one of them. It
stands not far from the banks of the Lymington
river, on the left bank as you look towards the
sea, between Roydon Farm and Boldre Church.
It occupies a small knoll, surrounded with other
trees, some of which are Yews, but of inferior
beauty. A little stream washes the base of the
knoll, and winding round forms it into a penin-
sula. If any one should have the curiosity to
visit it, from this description, and by the help of
these landmarks, I doubt not he may find it, at
any time within the space of these two or three
centuries, in great perfection, if it suffer no external injury. If such trees were common, they
would recover the character of the Yew-tree among
the admirers of picturesque beauty."

Sir Walter Scott is of a different opinion:—

> " But here, 'twixt rock and river, grew
> A dismal grove of sable hue,
> With whose sad tints were mingled seen
> The blighted Fir's sepulchral green ;
> Seemed that the trees their shadows cast
> The earth that nourished them to blast,
> For never knew that swarthy grove
> The verdant hue that fairies love ;
> Nor wilding green, nor woodland flower,
> Arose within its baleful bower ;
> The dank and sable earth receives
> Its only carpet from the leaves,
> That, from the withering branches cast,
> Bestrew'd the ground with every blast."

The wood of the Yew, Loudon says, is hard, compact, of a fine and close grain, flexible, elastic, splitting readily, and incorruptible. It is of a fine orange red, or deep brown; and the sap wood, which does not extend to a very great depth, is white, and also very hard. The fineness of its grain is owing to the thinness of its annual layers, two hundred and eighty of these being sometimes found in a piece not more than twenty inches in diameter. " The Yew was formerly what the Oak is now, the basis of our strength. Of it the old English yeoman made his long-bow, which, he vaunted, nobody but an Englishman could bend. In shooting he did not, as in other nations, keep his left hand steady, and draw his bow with his right; but keeping his right at rest upon the nerve, he pressed the whole weight of his body into the horns of his bow. Hence probably arose the English phrase of *bending* a bow, and the French of *drawing* one. Nor is the Yew celebrated only for its toughness and elasticity, but also for its durable nature. Where your paling is most exposed either to winds or springs, strengthen it with a post of old Yew. That hardy veteran fears neither storms above nor damps below. It is a common saying amongst the inhabitants of New Forest, that a post of Yew will outlast a post of iron."*

The Yew is propagated either by seeds or by cuttings, but in whatever way young plants are reared, they grow very slowly.

A variety, called the Irish Yew, is cultivated, which has upright branches, and leaves which are not arranged in opposite ranks, but scattered on all sides of the stem. It is very plentiful near Antrim, where there are said to be specimens more than a century old. Another variety is found in the grounds of Clontarf Castle, which differs from the common kind in bearing yellow berries.

Of the Yew there are a greater number of remarkable specimens than of any other British tree. Among these the first which deserve notice are the following, mentioned by Evelyn.

" A Yew-tree in the churchyard of Crowhurst, in the county of Surrey, which I am told is ten yards in compass, but especially that superannuated Yew-tree growing now in Braburne churchyard, not far from Scot's Hall in Kent; which being fifty-eight feet eleven inches in the circumference, will be about twenty feet diameter, as it was measured first by myself imperfectly, and then more exactly for me, by order of the late Right Honourable Sir George Carteret; not to mention the goodly planks, and other considerable pieces of squared and clear timber, which I observed to lie about it, that had been hewed and

* Gilpin's " Forest Scenery."

sawn out of some of the arms, only torn from it by impetuous winds." The former of these trees is still in existence; it stands near the east end of the church, and measures ten yards nine inches in girth, at about the height of five feet from the ground. The trunk has been hollowed out, and is filled up with narrow benches round the sides, and a table in the centre. At a farm-house hard by is shewn a cannon-ball weighing upwards of twenty-five pounds, which is said to have been found imbedded in the trunk when it was hollowed out many years ago. The tree is dead at the top, and the branches to the south were broken off by a violent storm on the 22nd of December, 1845. It consequently has a very naked appearance on this side; but the remaining branches, though much decayed, are still green and healthy.* The figure at the head of this chapter represents the tree as it appeared previously to the storm of 1845.

The Rev. James Brothers, Vicar of Brabourne, informs me that the second of the trees recorded by Evelyn has so long ceased to exist, that not the least traditionary remembrance of it remains in the parish or neighbourhood. Tourists, misled by the Histories of Kent, which describe the tree as if it yet stood, pay frequent visits to Brabourne, and are not a little disappointed at not being able to find even a relic of the venerable monster.

" In the churchyard of the village of Gresford, Denbighshire, stands a Yew-tree, which measures thirty feet in girth at the height of four feet from

---

* This account was kindly furnished me by the Rev. Fitzherbert Fuller.

the base; the branches are in themselves large trees, and shade the ground to a great extent. It must be of incalculable antiquity, and it is not yet in a state of much decay, though it has perhaps for centuries attained its maturity. Other Yew-trees, in the same place, which were planted, as the parish register records, in the year 1727, are, on an average, somewhat more than four feet in girth."*

" At Crom Castle, the seat of the Earl of Erne, in the county of Fermanagh, Ireland, in the garden of the old castle, which was burnt down

YEW-TREE AT CROM.

about a century ago, stands the most celebrated Yew-tree in Ireland. I have not been able to arrive at anything like certainty as to its age. Old men of eighty years remember it in their boyhood as the old Yew-tree of Crom. The appearance of the tree is that of an enormous green mushroom. The stem rises from a small mound, and is not above eight feet in height, and about three feet in diameter. The branches spread in great numbers horizontally from the trunk, and are supported on a number of wooden pillars, with gravel walks between them. These branches

* " Gardeners' Chronicle."

extend over a space of about seventy-five feet in diameter; the height of the tree is about twenty-one feet. A party of two hundred have often dined under this tree."[*]

The Great Yew at Ankerwyke, near Staines, which is described and figured by Strutt, is supposed to have flourished upwards of a thousand years. It measures twenty-seven feet eight inches in circumference at three feet from the ground; at eight feet, thirty-two feet five inches. Immediately above the latter height, there are five principal branches, each of which is upwards of five feet in circumference. Above these branches, the trunk measures twenty feet eight inches. Its height is forty-nine feet six inches, and the diameter of its shade sixty-nine feet. It is noted not only for its antiquity and immense size, but as the place of conference of the barons who compelled King John to sign Magna Charta at Runnymede, in its immediate vicinity. Tradition says also, that it was the favourite meeting-place of Henry VIII. and Anne Boleyn.

The Yew-trees at Fountain's Abbey are said to have afforded shelter to the monks while their Abbey was being built in 1132. They were formerly seven in number; but the largest was blown down about the middle of the last century. They are of an extraordinary size; the trunk of one of them is twenty-six feet six inches in circumference, at the height of three feet from the ground, and they stand so near each other as to form a cover almost equal to a thatched roof. They are described by Strutt as being highly picturesque.

* "Gardeners' Chronicle."

The Mamhilad Yew, described and figured by Loudon, stands in the churchyard of Mamhilad near Pontypool. It is twenty-nine feet in circumference, and at four feet from the ground divides into six main boughs. The trunk is hollow, and on the north side has an opening down to the ground, through which is seen another and apparently detached Yew, several feet in diameter, covered with bark, and in a state of vigorous growth; it is, in fact, a great tree, and overtops the old one.

The Fortingal Yew stands in the churchyard of Fortingal, or the Fort of the Strangers, in the heart of the Grampians, and is in all probability the oldest tree in Scotland. This prodigious tree was measured by Judge Barrington before the year 1770, and is stated by him to have been at that time fifty-two feet in circumference; when measured by Pennant some years afterwards, it was fifty-six feet six inches in circumference. Persons then alive remembered the time when its trunk was continuous to the height of three feet, but it is now decayed to the ground, and completely divided into two distinct stems, between which, funeral processions formerly used to pass. Dr. Neill says, that since 1770 it has been much injured by the removal of large branches, carried off by the country people for the sake of being made into cups and other relics, which visitors were in the habit of purchasing. What existed of the trunk in 1833 presented the appearance of a semicircular wall, exclusive of the remains of some decayed portions, which scarcely rose above the ground. It is now surrounded by an iron rail, and this venerable tree, which in all

FORTINGAL YEW.

probability was in a flourishing state at the commencement of the Christian era, may yet survive
for centuries to come. In Strutt's "Sylva Britannica" there is an admirable figure of the Fortingal Yew, of which the accompanying woodcut
is a copy.

The Ribbesford Yew is remarkable rather for
the singular nursery in which it chose to establish
itself than for great size. The seed from which
this tree sprung, must have been dropped, probably by a bird, on the summit of a Pollard Oak,
seventeen feet in circumference. As the Oak
decayed, the roots of the Yew gradually penetrated
downwards until they reached the ground, and
being eventually converted into stem filled up
the whole cavity of the trunk. Previously to the
year 1845, both the Yew and the Oak had numerous spreading branches, which mixed their
foliage in a very pleasing manner. A writer in the
"Analyst" stated his opinion, that it would in a
few years increase to such a size as to burst asun-
the oaken shell which enclosed it and ultimately
stand alone, as if it had sprung up from the
ground. In the year above mentioned the event
which had been anticipated, took place, for I am
informed by the Rev. E. W. Ingram that a violent
hurricane in July divested the Yew of the surrounding Oak, and that no vestiges of its foster
parent now remain.

Several instances are on record of old Yews
being renewed by a singular natural process.
When the upper part of a trunk begins to decay,
the crumbling wood forms a rich soil, into which
a young shoot from a neighbouring bough sends
a root. The young branch thus nourished inde-

pendently of the old roots, grows vigorously, and
in time becomes a tree, standing in the centre of
the hollow trunk, remaining perhaps partially
united to the parent, but deriving its support
principally from the soil. A tree thus formed,
several feet in diameter, stands in the centre of
the great Yew at Mamhilad (described above),
and will probably continue to flourish for centuries after the wooden walls with which it is
enclosed have crumbled to dust. A similar phenomenon has been observed in the Willow.

THE FIR TRIBE.

CONIFERÆ.

Class—MONŒCIA.   *Order*—MONADELPHIA.

THE trees of this Order seem, from their structure and habit, to be specially designed to occupy
stations which are, in more than an ordinary
degree, exposed to the violence of wind and
weather. Accordingly we find all the species,
with very few exceptions, flourishing in extreme
magnificence on the mountains of the cold
and temperate regions of the earth; but, even
when planted on the lowlands, they retain their
characters so constantly, that we can at a glance
distinguish them from any other trees with
which they may be associated, whether deciduous
or evergreen. The mountains are their natural
haunts, but some of them will flourish with
tolerable luxuriance in other situations, while
others, like human mountaineers torn from their
beloved Alpine homes, dwindle away and soon
perish, their very decay being accelerated by the
nursing and pruning and other means adopted to
promote their welfare.

The principal characters by which the Fir-tribe
are fitted for their native haunts are these :—
Springing from the bare crags, or a stratum of
dry soil, which is incapable of affording nourishment to any moderately sized plant furnished
with roots having a downward tendency, the Firs,

both young and old, extend their roots horizontally, or in a direction parallel to the surface of
the ground,—tap-root they have none, for such
an appendage would be useless to trees often growing in soil but a few inches deep. The roots
being, moreover, close to the surface, or even
partially above it, acquire a hardness and toughness which enable them to resist the action
of the wind on the head of the tree much more
effectually than in the case of trees whose
juicy roots run deep into the ground. It is a
well-ascertained fact in physics, that any given
number of separate strings will support a much
heavier weight than if they were united into one
rope. This fact might have been inferred from
the roots of the Fir-tribe, for being required to
resist a greater degree of force than the roots
of other trees, they are smaller, and proportionally more numerous, thus being stronger in themselves, and presenting a larger surface of resistance to the surrounding soil, that is to say, being
both less liable to snap, and to be torn out from
the ground. Moreover, as they extend in all
directions, they are prepared to resist the violence
of the mountain tempest, no matter from what
quarter it may proceed.

From the centre of this web of wiry roots
rises a stiff column of solid timber, the strength
of which is not impaired by being divided into
branching arms, but the whole substance is
thrown into one trunk, so that here the least
possible amount of surface is exposed to the
action of the wind. The Firs, too, are eminently social trees; it rarely happening in the
mountains that one stands alone: but though

social among themselves, they are strictly exclusive as it regards other trees; they are generally
found covering extensive tracts of country, and
being evergreen, they shut out the light from
every other tree that attempts to germinate
beneath their unfriendly shade. For the most
part, they stand as close together as is consistent
with their healthy growth; hence they not only
borrow from each other a firmer hold of the
ground by interlacing their roots, but prevent a
free circulation of air round their stems, and
consequently the small lateral branches which are
formed, soon grow sickly and perish. This
effect is perhaps increased by the rarified state of
the air at great elevations. The decay of the
lateral branches does not, however, in any case,
extend to the bole, for the dead wood of the
Firs does not rot, as is the case in other trees, but
" as soon as vegetation ceases, the consistence of the
wood changes; the sap disappears, and the wood,
already impregnated with resinous juice, becomes
surcharged to such a degree as to double its
weight in a year."* Meanwhile the trunk in-
increases in dimensions, and encloses the hard
stump in its substance; and hence originate the
dark circular knots, so common in most kinds of
Fir-wood. In the main stem the woody fibres
are less close than in most other trees; the effect
of which is, that the wood is more elastic,
bending before the blast, but not breaking; and
the resinous nature of the juices in every part of
the tree defies the influence of the severest
frost.

On examining the leaves, we find an equally

* Michaux.

beautiful adaptation of these organs to the circumstances in which they are placed. The thin
dilated leaves commonly to be found during the
summer months on deciduous trees, in the
plains, would here be soon torn to pieces or scattered by the wind; if, on the other hand, they
partook of the character of the lowland evergreens, such as the Laurel and Bay, that is to say,
if they had a broad surface and a tough substance,
the very resistance they offered would bring
destruction on the tree they clothed. The wind
would act on them mechanically, like the force
exerted on the long arm of a lever, and the
breeze, instead of passing freely through the
branches with a low murmur (one of the plea-
santest sounds in nature), would be as destructive
as the most terrific hurricanes which occasionally devastate the forests of countries within the
Tropics.

But besides being admirably adapted for with-
standing the violence of the storms, to which
the Firs are, from their situation, peculiarly liable,
the leaves of these trees are no less remarkable in
other respects. Subject to almost uninterrupted
exposure to cold, the resinous juices in which
they abound serve as a safeguard against its
injurious effects, and yet their shape is such as to
be naturally the cause of their temperature being
lower than that of surrounding bodies. A person
walking through a mist will soon find his eye-
lashes and hair covered with small drops of water
while the rest of his person remains dry; if he
examines the ground, he will also find that the
blades of grass by the way-side are fringed with
dew-drops, while the road itself is quite free from

moisture. This phenomenon is owing to the excessive radiation of heat from bodies which present a large surface in proportion to their bulk, and the consequent condensation of moisture on cold substances. Precisely the same effect is produced on the leaves of the Firs, which are therefore said, though inaccurately, to attract moisture; the true state of the case being that they are reduced to a low temperature by excessive radiation of heat into space, and are consequently subject to a constant deposition of moisture in the shape of clear globules, which soon becoming too heavy to remain suspended on the leaves, fall to the ground, and, having supplied the scanty soil with a sufficiency of nourishment for the thirsty roots, trickle away in little rills. These either sink into the ground, and reappear below as mineral springs, or flow along the surface, continually increasing from the accession of similar tributaries, and fertilizing the valleys through which the very same mists had previously been carried, where they had encountered no substances of a temperature low enough to arrest their progress.

Again, if the leaves of mountain trees grew in tufts at the extremities of the boughs, and were dilated like those of the trees alluded to above, they would, whenever a fall of snow occurred (and this must frequently be the case) afford a convenient resting place for the descending flakes, which would soon accumulate, and form a mass too heavy for the branches to support, and every such occurrence would spread ruin through the forest. But slender as they are, and rarely or never horizontal in their direction, they do not allow the snow to accumulate on the extremities. Often indeed the tops of the trees retain their sombre hue, while the ground is thickly covered with a white mantle; but though they refused to give the fleecy particles a lodgment above, they do their utmost to retain it below, protracting the period of thaw-ing by intercepting the sun's rays, as well as by re-ducing the temperature of the air around them by abundant radiation. We consequently find the ground in woods covered with a sheet of snow long after every particle has been thawed in the out-skirts of the forest, and the soil yet moist beneath the shade, when the exposed country is parched with drought.

BUDS OF STONE-PINE.

There is yet another peculiarity of the Fir-tribe connected with this subject which deserves notice. The perfection of the Fir, as has been already noticed, consists in height rather than lateral expansion. In all other trees (except the Palms) a bud is produced in the axil of every leaf. This is not the case in the Firs, but buds are produced very sparingly, and nearly always at the extremities of the shoots. Provision is thus made for the upward growth of the tree, but not for its lateral expansion. In other trees, again, the unfolding of all the buds on an individual is simultaneous, or nearly so; but in the case of the Fir-tribe, " the bud which terminates the summit of the tree, and is destined to form its leading shoot, and increase its height, is developed the last; and this delay seems a provision of nature for the safety of the most important shoot which the tree can produce ; thus insuring its height rather than its breadth, and the production of timber by the preservation of its permanent trunk, rather than of its temporary and comparatively useless branches."*

It might be supposed that the Firs, exposed as they are to the action of the most violent thunderstorms, would be liable to be shattered by discharges of the electric fluid to an extent not known in the case of any other trees. The reverse of this is the case ; for they are furnished with a natural apparatus, which not only in most cases protects themselves from the effects of lightning, but tends to equalize the electric condition of the atmosphere, and so to extend their influence to districts indefinitely remote. Fresh vegetables in general conduct the electric fluid with facility, owing to the good conducting properties of the

* Loudon.

fluids which they contain. If a small blade of
grass be placed in contact with the conductor of
a powerful electrical machine in operation, the
whole of the electricity will be found to be
carried off by the blade of grass. Pointed conductors, and especially vegetable conductors, are
admirably fitted to receive and disperse electricity,
it having been found by experiment that a few
blades of grass placed near the brass knob at the
top of a Leyden jar, will quickly and silently
discharge it. It has been found impossible
to give an electric shock to a circle of people standing on a lawn, as the electricity took
the shorter and better conducting course through
the grass; and it has also been found, that
when the electroscope, (an instrument for measuring the degree of electricity,) indicated abundance of electricity in the free open air, it
indicated none in the vicinity of a tree with
pointed leaves. It is not unfair, therefore, to
assume that every one of the myriads of pointed
conductors in the Pine forests of Norway and
Russia, is continually employed in withdrawing
electricity from the atmosphere, and contributing
to promote an equable electrical condition in the
atmosphere of places far remote.

The flowers of the Pine are of two kinds, both
of which are of a simple structure, being destitute both of calyx and corolla, and therefore not
liable to be torn by the wind. The barren
flowers are scaly catkins, and contain an unusual
quantity of pollen, which is sometimes carried
away by storms, and descends in remote districts,
in the shape of clouds of sulphur-coloured dust, to
the great terror of the superstitious. The fertile

flower is a solid catkin, composed of thick over-
lapping scales, at the base of each of which are
generally two ovaries. The whole of the fertile
flower is persistent, increasing in size, but not
altering materially in shape until it becomes
a woody cone. Meanwhile the ovaries have grown
into seeds, furnished each with a membranous
wing, which, though not buoyant like the down
of the thistle, flies away lightly enough before the

CONES OF STONE-PINE.

mountain breeze. Until the seeds are ripe, that
is, for a year or more after flowering, the cones
are hard balls of wood, composed of a number of
distinct pieces, so closely ad-
hering together, that not a
drop of water can penetrate
them, and firm enough to bear
the shock of dropping from
the loftiest trees, or of leaping from rock to rock

SEED OF SCOTCH PINE.

without injury. When the seeds are thoroughly ripe, but not before, the cones, whether remaining attached to the tree or lying on the ground, open spontaneously, and allow the seeds to escape.

Thus a constant succession of young plants is kept up, a provision which, in the case of this tribe, is the more necessary from the fact that they send up no suckers from the roots, and when cut or blown down they never send up new shoots from the mutilated trunk. Their duration, too, in most instances, is less than that of other forest-trees.

Seedling Firs are remarkable for being composed of five or six seed-leaves, which in their youngest stage are united at their points by the shell of the seed. When this falls off they

SEEDLING.

spread, and a bud containing true leaves rises from the centre.

The geographical range of the Fir-tribe is extensive, but they are most abundant in the temperate parts of the northern hemisphere. Some species are found both in Europe and America, so far north as to border on the regions of perpetual snow; and others, in central Europe and in Asia, on the Alpine and Himalayan mountains, where, from their great elevation, the climate is equally cold. Other species occupy the same position in the mountains of America, extending to the height of more than twelve thousand feet, beyond which altitude vegetation entirely ceases.

Frequent mention occurs in the Sacred Writings of the Cedar and Fir, the wood of both trees being peculiarly adapted for building. The Cedar still flourishes on the same site which it occupied in the days of Solomon; the Fir (Beroth) is supposed to be the same with the Cupressus of the Latins and our Cypress, a common tree in the East. Solomon employed both Cedar and Fir in the erection of the Temple, the floor of which was of Fir; the musical instruments of David were of the same wood. Pliny mentions that the doors and other parts of the temple of Diana, at Ephesus, were made of Cypress-wood. The Thyine-wood (Rev. xviii. 12) is supposed to be another species of Cypress. The wealthy among the Romans adorned their villas with this wood; Varro, describing the splendour of a certain villa, celebrates the golden decorations, but praises in still higher terms the wainscoting of precious Thyia-wood. Being so much in demand, it became an important article of trade, and is therefore classed among the precious merchandize of fallen Babylon. The Gopher-wood, of which the Ark was built, is thought to be another species of Cypress: being at once light, and not subject to rot, it was often used in ship-building. Alexander the Great caused the great fleet which he prepared, to be constructed of Cypress wood, which was brought from Assyria.

The Talmudists relate, that it was customary in Judæa for each family to plant a Cedar before the house at the birth of a son, and a Fir at the birth a daughter. These trees were deemed sacred, and were not cut down till the children were grown up and needed the timber for their

household furniture. At the time when Judæa
was subject to the Romans, after the destruction
of Jerusalem by Titus, the daughter of the Emperor Adrian happened to be travelling through
that country, when her chariot was injured, and
her attendants proceeded, in an overbearing manner, to cut down one of the sacred trees, to be
used in repairing it.   The inhabitants of the
place rose and massacred the train of the princess,
who was so enraged that she forced her father to
make war against the Jews, to humble their
pride.

Herodotus tells us, that Miltiades, at the head
of the Thracian Dolonci, having made war on
the people of Lampsacus, was taken prisoner by
an ambuscade. His friend, Crœsus, having heard
of his misfortune, sent a herald to the Lampsacans, threatening them that unless they released their prisoner, he would cut them down
like a Fir-tree. The Lampsacans were at first
perplexed, but when one of their wise men
reminded them that the Fir-tree, if once cut
down, never shoots again, they were so terrified,
that they dismissed their prisoner forthwith.

The victors at the Isthmian games, held at
Corinth, were crowned with garlands of Pine-
branches. The cones were used by the Romans
to flavour their wines, being thrown into the
vats, and suffered to float,—a custom which is still
in existence in Italy. Hence the thyrsus, or
wand of Bacchus, terminates in a Fir-cone. The
timber was employed by both Greeks and Romans
in naval and domestic architecture, and the
various resinous productions were extracted by a
method very similar to those now adopted. The

Pine appears to have been held sacred by the Assyrians. Mr. Layard informs us that on the sculptures discovered by him during his excavations at Nimroud, the ancient Nineveh, there are many representations of figures bearing a Fircone.

Descriptions of the most remarkable species of Fir introduced into Britain (for we can claim one only as a native) will be given under their several heads. The following general remarks will, nevertheless, be read with interest:—

" If the reader will cast his eyes on the map of Sweden, and imagine the Gulf of Bothnia to be surrounded by one continuous unbroken forest, as ancient as the world, consisting principally of Pine-trees, with a few mingling specimens of Beech and Juniper, he will have a general and literally correct notion of the real appearance of the country."

" As we proceeded to Hamrange, we passed through noble avenues of trees, and saw some fine lakes on either side of the road. Some of the forests had been burned, by which the land was cleared for cultivation. The burning of a forest is a very common event in this country; but it is more frequent towards the north of the Gulf of Bothnia. Sometimes a considerable part of the horizon glares with a fiery redness, owing to the conflagration of a whole district, which, for many leagues in extent, has been rendered a prey to the devouring flames."*

" Sometimes the monotony of the Pine barren was interrupted in no very pleasant style by the heat and smoke arising from the forest being on

* Dr. Clarke.

fire on both sides of us, though, as it happened, we were never exposed to any danger, or to serious inconvenience, in consequence of these conflagrations. The tree in the foreground had caught fire near the ground, and having, I do not know how, been hollowed out in its centre, the flames had crept up and burst out some feet higher, so that they were roaring like a blast-furnace, and rapidly demolishing the tree at the bottom, while the branches at the top were waving about in full verdure, as if nothing unusual was going on below."*

Linnæus well describes the danger by which he was surrounded when traversing one of these burning forests in Lapland. "Several days ago the forests had been set on fire by lightning, and the flames raged at this time with great violence, owing to the drought of the season. In many different places, perhaps nine or ten that came under my notice, the devastation extended several miles' distance. I traversed a space of three quarters of a mile in extent (about four miles and a half English), which was entirely burnt, so that, Flora, instead of appearing in her gay and verdant attire, was in deep sable,—a spectacle more abhorrent to my feelings than to see her clad in the white livery of winter; for this, though it destroys the herbage, leaves the roots in safety, which the fire does not. The fire was nearly extinguished in most of the spots we visited, except in ant-hills and dry trunks of trees. After we had travelled about half a quarter of a mile across one of these scenes of desolation, the wind began to blow with rather more

* Hall's " Sketches in Canada."

force than it had previously done, upon which
a sudden noise arose in the half-burnt forest, such
as I can only compare to what may be imagined
among a large army attacked by an enemy. We
knew not whither to turn our steps; the smoke
would not suffer us to remain where we were,
and we durst not turn back. It seemed best
to hasten forward in hopes of speedily reaching
the outskirts of the wood, but in this we were
disappointed. We ran as fast as we could, in
order to avoid being crushed by the falling trees,
some of which threatened us every minute.
Sometimes the fall of a large trunk was so sudden
that we stood aghast, not knowing which way
to turn to escape destruction, and throwing ourselves entirely on the protection of Providence.
In one instance a large tree fell exactly between
me and my guide, who walked not more than a
fathom from me; but, thanks to God, we both
escaped in safety. We were not a little rejoiced
when this perilous adventure terminated, for we
had felt all the while like a couple of outlaws in
momentary fear of surprise."*

The burning of these forests, however, is in-
correctly attributed to the effects of lightning.
Fires of this kind have been traced to the careless-
ness of the Laplanders and boatmen on the rivers,
who, using German tinder to light their pipes,
suffer it to fall in an ignited state among the dry
leaves and moss. They also leave large fires
burning, which they have kindled in the midst
of the woods to drive away the mosquitoes; and
in either of these ways the fire is easily communicated to the surrounding trees.

* Lachesis Lapponica.

It has been observed above that the Firs are
less liable to be overladen with snow than if they
were furnished with broad leaves, like the trees
which grow in the plains. The Norway Spruce,
it will be seen hereafter, is more subject to acci-
dents from this cause than other European species;
and when, as sometimes happens, the overcharged
clouds descend in the form of rain, which freezes
fast to the leaves and branches, the effect is
terrific.

A traveller in the Alleghany mountains relates,
that on the morning after an "ice-storm" of this
kind, "the accumulation of ice on the branches
of the forest trees presented a beautiful and
extraordinary spectacle.   The noblest timber-
trees were everywhere to be seen bending beneath
the enormous load of ice with which their branches
were encrusted, and the heavy icicles which
thickly depended from every point,—the thickness
of the ice, even on the spray, often exceeding an
inch. The smaller trees, from twenty feet to
even fifty feet in height, were bent to the ground
by this unwonted burden, and lay pressing on
one another, resembling fields of gigantic corn
beaten down by a tempest. Above, the taller
trees drooped and swung heavily : their branches
glittering, as if formed of solid crystals ; and,
with the slightest breath of wind, clashing against
each other, and sending down showers of ice.
The following day, the limbs of the trees began
to give way beneath their load. The leafy spray
of the Hemlock Spruce was thickly incased, and
hung drooping round the trunks upon the long
pliant branches, until the trees appeared like
solid masses or monumental pillars of ice. Every

where around was heard the crashing of the
branches of the loftiest trees of the forest, which
fell to the earth with a noise like the breaking
of glass, yet so loud as to make the woods resound. As the day advanced, instead of branches,
whole trees began to fall, and during twenty-
four hours the scene which took place was as
sublime as can well be conceived. There was
no wind perceptible, yet, notwithstanding the
calmness of the day, the whole forest seemed
in motion, falling, wasting, or crumbling, as
it were, piecemeal. Crash succeeded to crash,
until at length these became so rapidly continuous as to resemble the incessant discharges
of artillery, gradually increasing, as from the
irregular firing at intervals of the outposts, to
the uninterrupted roar of a heavy cannonade.
Pieces of a hundred and fifty, and a hundred
and eighty feet in height, came thundering to
the ground, carrying others before them. Under
every tree was a rapidly accumulating *débris* of
displaced limbs and branches; their weight increased more than tenfold by the ice, and crush-
ing everything in their fall, with sudden and
terrific violence. Altogether, this spectacle was
one of indescribable grandeur. The roar, the
cracking and rending, the thundering fall of
the uprooted trees, the startling unusual sounds
and sights produced by the descent of such
masses of solid ice, and the suddenness of the
crash, when a neighbouring tree gave way,
was awful in the extreme. Yet all this was
going on in a dead calm, except, at intervals,
a gentle air from the south-east slightly waved
the topmost pines. Had the wind freshened,

the destruction would have been still more appalling."*

The manner of conveying Fir-timber from these forests to the sea is thus described by Dr. Clarke : " At Helsinborg some Fir-trees of astonishing height were conducted by wheel-axles to the water side. A separate vehicle was employed for each tree, drawn by horses which were driven by women. These long, white, and taper shafts of deal timber, divested of their bark, afforded the first specimens of the produce of those boundless forests of which we had, till then, formed no conception."

" We remember," says Sir T. D. Lauder, " having been thrown into some degree of alarm by encountering one of the enormous Silver Fir-logs, as we were going up the Jura. Our calèche had got three parts of the way up the hill, where the road gradually ascended along the perpendicular face of the mountain, and hung over the valley and village we had left, which looked like some toy as it lay in the narrow bottom below, when we were suddenly met by a vast Pine-log, laid, not on two carts, as with us, to keep it steady, but on one cart, whence both the butt and top projected, so that the point jerked backwards and forwards with terrific force and elasticity.

" To our infinite horror this magnificent sylvan specimen took the wall of us, and we must confess that we have spent few more anxious minutes during our lives than that which was occupied in steering by this danger ; for if we had been but touched by the tree in its jerking

* R. C. Taylor.

vibrations, the carriage, horses and all, would have been launched into the air, exactly as a small beetle might be fillipped by the finger from the sill of a garret window into the street."

In the forest districts of the Alps, of Germany, and of Norway, where the people derive a good part of their existence from the timber of their trees, the modes of transporting the produce to the markets are often highly curious. In some cases the woodmen cut down the trees, hurl or roll them into a mountain stream, and let them float down to the sea, or a lake, or to any place where they can be conveniently disposed of. This is comparatively easy so long as the forest is not far from a stream; but when it is inland, or situated at a great height, or separated from a stream by a rugged and mountainous district, the ingenuity of the woodman is taxed to the utmost to devise means of transporting the timber. One of the means adopted is to construct a slide, down which the trunk may run by its own impetus. Early in spring the woodmen set off, to begin their business of cutting down the trees in the forest, perhaps many miles from their homes; they have to construct rude huts, in which they live during the summer and autumnal months, and throughout the whole of this period they employ themselves in cutting down the noble trees which surround them. Every tree is classed according to its fitness for practical purposes, and cut up into logs; and the logs so accumulated are heaped up into huge piles. When the winter arrives, all these logs are transported down to some stream or lake, by means of a slide or trough. This trough is usually constructed of six or eight fir-trees, placed side by side lengthways, so as to form a semicircular channel, made smooth by stripping the bark from the trees. The trees are laid side by side, and end to end, till the slide is of considerable length, having a gradual descent, curving round the shoulders of mountains, spanning over valleys and yawning ravines by means of viaducts, and even perforating solid rocks by means of tunnels. In the year 1810, when the price of Baltic timber had attained its greatest height, a stupendous, and at the same time successful, effort was made to convey the timber of Mount Pilate to the Lake of Lucerne, whence it might be floated down the Rhine to the sea. Under the superintendence of M. Rupp, a slide was constructed, six feet broad, and from three to six feet deep, and extending to a distance of forty-four thousand feet (eight miles). It was completed in 1812, and twenty-five thousand Pine-trees were employed in its construction. It was called the slide of Alpnach, from the name of a village near it. The logs were drawn to the trough either by hand-sledges or by oxen, and placed in it at the top; the snow was partially cleared away from the trough, and a few logs were thrown to clear the channel. Water was next poured upon it, which quickly froze, forming a surface of ice through its entire extent. The logs placed on the upper surface of this slippery trough immediately descended, slowly at first, but with almost inconceivable velocity, as their momentum increased. When the operations were to begin, workmen were posted at regular distances; and as soon as everything was ready, the workman at the lower end of the slide cried out to the one above him, " *lachez*," (let go). The cry was repeated from one to another, and reached the top of the slide in three minutes. The workman at the top then cried out to the one below him, " *il vient* " (it comes) ; and the tree was instantly launched down the slide, preceded by the cry, which was repeated from post to post. As soon as the tree had reached the bottom, and plunged into the lake, the cry of " *lachez* " was repeated as before, and a new tree was launched in a similar manner. By these means a tree descended every five or six minutes. The velocity with which the trees descended is almost inconceivable ; the descent of eight miles was usually made in six minutes, but in wet weather it was frequently effected in three, being at the rate of a hundred and eighty miles an hour ! Perhaps the best way of conveying an idea of this amazing velocity is to state, that a spectator standing by found it quite impossible to give two successive strokes with his stick to any, even the longest, tree, as it passed him. The logs entered the lake with so much force that many of them seemed to penetrate its waters to the very bottom. Much of the timber of Mount Pilate was thus brought to market ; but the expense attending the process rendered it impossible for the speculator to undersell the Baltic merchant, after the arrival of peace had opened the market for his timber, and so the slide of Alpnach fell into ruin.

An interesting description has been given by Howison of the mode of bringing timber to market in the heart of Russia. A Russian proprietor who wishes to dispose of the timber on his property, having completed a bargain with a St. Petersburg merchant, sets his peasantry to work in selecting and felling the trees, and dragging them from the forests to the lakes and rivers. This work usually takes place during the winter months, in order that everything may be ready for floating the timber to the capital as soon as the ice in the rivers and lakes breaks up. As the ground is generally covered several feet deep with snow, and as the trees judged to be sufficiently sound and large for the market lie widely apart, the workmen employed in selecting them are compelled to wear snow-shoes to prevent themselves from sinking in the snow. When the trees are found, they are cut down with hatchets, and the heads and branches lopped off. The trunk is then stripped of its bark, and a circular notch is cut round the narrow end of it, to facilitate the fixing of the rope by which the horses are to drag the trunk along; and a hole is made in the other end to receive a handspike to steer the log over the many obstacles that lie in its way. Many of these trees are seventy feet in length, and of proportionate diameter; and they are drawn by from four to nine horses each, yoked in a straight line, one before another, since the intricate narrow paths in the woods will not permit any other arrangement. One man mounts upon the leading horse, and another upon the middle one, while others support and guide with handspikes the large and distant end of the tree, to raise it over the elevations of snow, and make it glide smoothly along. The conveyance of these large trees, the long line of horses, and the number of

peasants accompanying them through the forest,
present a very picturesque appearance. In many
cases the trees are brought nearly a thousand
miles before they are delivered to the merchant;
and they generally remain under his care till
another winter, to be shaped and fitted for exportation in such a manner as to take up as little
room as possible on shipboard; so that this
timber does not reach the foreign consumer till
two years after it has been cut down. When the
trees are delivered to the merchant, they are carefully examined to ascertain their soundness; and
for this purpose a hatchet is struck several times
against them, the emitted sound affording the
means of estimating the soundness of the tree:
those which are defective constitute about one
tenth of the whole. The trees are not conveyed
from the forests the whole way to St. Petersburg
by horses, but only to the margin of some stream
or lake, from whence they may be floated down to
the capital.

"The most striking examples of the floating of
timber by rafts are presented on the Danube and
Rhine. The immense forests of southern and
western Germany are in most cases within reach
of some stream or other which flows into the
Rhine, the Danube, the Rhone, or one of the
other large rivers; and in such cases the logs of
timber, precipitated into the smaller streams by
the troughs, or by some other contrivance, are
floated singly down these small streams until they
reach the larger rivers, when they are made into
rafts. "Below the bridge at Plattning, on the
Danube, the raft-masters of Munich, who leave
that city every Monday for Vienna, unite their

rafts before they enter the Danube. They descend the Isar upon single rafts only, but upon reaching this point they lash them together in pairs; and in fleets of three, four, or six pairs, they set out for Vienna. A voyage is made pleasantly enough upon these floating islands, as they have all the advantages of a boat without the confinement. A very respectable promenade can be made from one end to the other, and two or three huts erected upon them afford shelter in bad weather, and repose at night."*

" A little below Andernach, the Rhine forms a small bay or inlet, where the pilots are accustomed to unite together the small rafts of timber floated down the tributary rivers, and to construct enormous rafts, which are floated down the Rhine to Holland, and there sold. These huge rafts have the appearance of floating villages, each composed of twelve or fifteen little huts, on a large platform of timber. The raft, which is frequently eight or nine hundred feet long, by sixty or seventy wide, is composed of several layers of timbers or trees placed one on another and tied together, the whole drawing about six or seven feet of water. Several smaller rafts are attached to the large one, besides a string of boats loaded with anchors and cables, and used for the purposes of sounding the river, and going on shore. The rowers and workmen sometimes amount to seven or eight hundred, superintended by pilots, and over the whole is placed a proprietor or manager, whose habitation is superior to the others. As the men live on board the raft, the arrangements for their comfort are very extensive.

* M. Planché.

Pigs, poultry, and other animals are kept on board, and butchers accompany the troop. A well-supplied boiler is at work night and day, in a kitchen built on the raft. The dinner-hour is announced by a basket stuck on a pole, at which signal the pilots give the word of command, and the workmen run from all quarters to receive their rations. The consumption of provisions is enormous; forty or fifty thousand pounds of bread, twenty thousand pounds of fresh meat, with a proportionate quantity of butter, salt meat, vegetables, &c., are demolished in the voyage from Andernach down to Holland."*

A large number of insects are described by Köllar which are destructive to the Fir-tribe, some by eating off the foliage, others by using the young shoots for their dwelling, others by feeding on the bark and wood, and others by attacking the roots of young trees. The most mischievous are the Pine Lappet Moth (*Bombyx Pini*), the Black-arch Moth (*Bombyx monacha*), and the Pine Saw-fly (*Tenthredo Pini*). The reader is referred to his " Treatise on Insects" for an account of these, the limits of the present work not allowing even an enumeration of the principal ones. A few only will be mentioned under the head of the particular tree which each infests.

* " An Autumn near the Rhine."

THE SCOTCH FIR.

THE SCOTCH FIR OR PINE.

PINUS SYLVESTRIS.

THE Scotch Fir is the only one which is a native of Britain. Julius Cæsar, it has been remarked above,* states that the Beech and the kind of Fir which was known to the Romans by the name of *Abies*, were not to be found in this island. With regard to the Beech, I have endeavoured to shew that he was in error; but in the other case he was probably correct, for the tree which the Romans called Abies does not appear to be the same with our Pine, but with what we call the Silver Fir, which was not introduced into England until the beginning of the seventeenth century. From remote antiquity, the Pine has grown in the Highlands of Scotland, and the occasional discovery of trunks of the same tree in peat-bogs sufficiently proves that it was at one time indigenous to England. Extensive and most magnificent forests of Pine still exist in Scotland, exhibiting a character which belongs to no British forests composed of other trees—so peculiar indeed, and so wild, that it would be almost as hardy to doubt their native origin, as to deny that the soil from which they spring is a constituent part of the country.

" In the forests of Invercauld and Brae-

* Vol. I. p. 313.

mar," says Sir T. D. Lauder, " the endless Fir
woods run up all the ramifications and subdi-
visions of the tributary valleys, cover the lower
elevations, climb the sides of the higher hills, and
even in many cases approach the very roots of the
giant mountains which tower over them ; yet with
all this, the reader is mistaken if he supposes that
any tiresome uniformity exists among these wilds.
Every movement we make exposes to our view
fresh objects of excitement, and discloses new
scenes produced by the infinite variety of the
surface.   At one time we find ourselves wander-
ing along some natural level under the deep and
sublime shade of the heavy Pine foliage, upheld
high over head by the tall and massive columnar
stems, which appear to form an endless colonnade ;
the ground dry as a floor beneath our footsteps,
the very sound of which is muffled by the thick
deposition of decayed spines with which the seasons of more than one century have strewed it ;
hardly conscious that the sun is up, save from the
fragrant resinous odour which its influence is
exhaling, and the continued hum of the clouds of
insects that are dancing in its beams over the tops
of the trees.   Anon, the ground begins to swell
into hillocks, and here and there the continuity of
shade is broken by a broad rush of light streaming down through some vacant space, and brightly
illuminating a single tree of huge dimensions and
of grand form, which, rising from a little knoll,
stands out in bold relief from the darker masses
behind it, where the shadows again sink deep and
fathomless among the red and grey stems ; whilst
Nature, luxuriating in the light that gladdens the
little glade, pours forth her richest Highland trea-

sures of purple heath-bells, and bright green bil-
berries, and trailing whortleberries, with tufts of
ferns and tall junipers irregularly intermingled.
And then, amidst the silence that prevails, the red
deer stag comes carelessly across the view, leading
his whole herd behind him; and, as his full eye
catches a glimpse of man, he halts, throws up his
royal head, snuffs up the gale, indignantly beats
the ground with his hoof, and then proudly moves
off with his troop amid the glistening boles.
Again the repose of the forest is interrupted by
the music of distant waters stealing upon the ear:
curiosity becomes alive, and we hurry forward,
with the sound growing upon us, till all at once
the roar and the white sheet of a cataract bursts
upon our astonished senses, as we find ourselves
suddenly and unexpectedly standing on the fear-
ful bank of some deep and rocky ravine, where the
river, pouring from above, precipitates itself into
a profound abyss, where it has to fight its way
through countless obstructions, in one continued
turmoil of foam, mist, and thunder.   The cliffs
themselves are shaken, and the pines quiver while
they wildly shoot, with strange and fantastic
wreathings, from the crevices in their sides, or
where, having gained some small portion of nutriment on their summits, they rear themselves up like
giants aspiring to scale the gates of heaven.   And
here, perhaps, a distant mountain-top may appear
over the deep green Fir-tops.   By and by, after
pursuing the windings of the wizard stream for
a considerable way upwards, we are conducted by
it into some wide plain, through which it comes
broadly flowing and sparkling among the opposing
stones, where the trees of all ages and growths

stand singly, or in groups, or in groves, as Nature
may have planted them, or the deer may have
allowed them to rise, where distant herds are
seen maintaining their free right of pasture—
where, on all sides, the steeps are clothed thick
with the portly denizens of the forest, and where
the view is bounded by a wider range of those
mountains of the Cairngorum group, which are
now ascertained to be the highest in Great Britain.   And finally, being perhaps led by our way-
ward fancy to quit this scene, we climb the rough
sides of some isolated hill, vainly expecting that
the exertion of but a few minutes will carry us to
its summit, that we see rising above all its woods.
And we do reach it—but not until we are toil-
worn and breathless, after scrambling for an hour
up the slippery and deceitful ascent.   Then what
a prospect opens to us, as we seat ourselves on
some bare rock!   The forest is seen stretching
away in all directions from our feet, mellowing as
it recedes into the farthest valleys amid the distant hills, climbing their bold sides, and scattering
off in detachments along their steeps, like the
light troops of some army skirmishing in the van;
and, above all, the bold and determined outlines
of Benmachdhuie, that king of British mountains,
and his attendant group of native Alps, sharply,
yet softly delineated against the sky, look down
with silent majesty on all below."

These mighty forests are indebted for their re-
newal to the membranous wings with which Pine-
seeds are furnished.   By help of these the seeds
are carried to a great distance by the violent
winds to which mountainous tracts are liable,
and everywhere find soil enough to supply their

slender wants. The rook, too, is one of Nature's planters of Pine woods. Forsyth* tells us that from the Highland forests there come clouds of rooks in search of food, sometimes in such heavy columns as to create alarm among farmers, as to where and on what they are to dine; and if it were not for the Pine, which yields them food as well as lodging, they would soon be called by dishonest names, which they would no doubt deserve. Yet of these clouds of rooks, as they fly high, and glide harmlessly over head on their homeward passage in autumn evenings, Scotland may be proud; for these sable birds have had their homes in the Highland glens time out of mind, and have sown the seeds of almost all the Fir-trees that are to be found in the natural forests. It is well known that the rook has a natural propensity to steal away to some lonely quiet place with its booty, such as a Fir-cone or a potato, and there to eat what he can, leaving the rest, which, in the case of the Pine-cone, is just what is necessary for the production of tim- ber; for the first heavy snow presses the shat- tered cone, with any seeds that may remain in it, close to the ground, and these seeds, finding themselves in good circumstances as to soil, mois- ture, and heat, soon vegetate in the open heath, and eventually become trees. Some of the rooks, it is said, do even more than this; they not only convey the cones to some lonely place, but take advantage of the workings of an under-ground quadruped as black as themselves, and may be sometimes seen actively employed in burying the cones in mole-hills.

* Gardeners' Chronicle.

" It is curious to observe," says Sir T. D. Lauder in another place, " how the work of renovation goes on in a Pine-forest. The young seedlings come up as thick as they do in the nurseryman's seed-beds; and in the same relative degree of thickness do they continue to grow, till they are old enough to be cut down. The competition which takes place between the adjacent individual plants creates a rivalry that increases their upward growth, whilst the exclusion of the air prevents the formation of lateral branches, or destroys them after they are formed. Thus Nature produces by far the most valuable timber; for it is tall, straight, of uniform diameter throughout its whole length, and free from knots; all which qualities combine to render it fit for spars, which fetch double or triple the sum per foot that other trees do. The large and spreading trees are on the outskirts of the masses, and straggle here and there in groups or single trees."

How little the hand of man has had to do at any period, except within the last fifty years, in planting the Pine in Scotland, appears from the numerous extensive tracts which were once crowded forests, but have been dismantled by human agency. Almost every district of the Highlands bears the trace of the vast forest with which, at no very distant period, the hills and heaths were covered : some indeed have decayed with age, but large tracts were purposely destroyed in the latter end of the sixteenth and beginning of the seventeenth centuries. On the south side of Ben Nevis, a large Pine-forest, which extended from the western Braes of Lochabar to the black water and mosses of Ranach, was

burned to expel the wolves. In the neighbourhood of Loch Sloy, a tract of woods, nearly twenty
miles in extent, was consumed for the same purpose ; and at a later period, a considerable part of
the forests adjoining Lochiel was laid waste by
the soldiers of Oliver Cromwell, in their attempts
to subdue the Clan Cameron. It is not above
eighty years since Glen Urcha was divested of a
superb forest of Firs some miles in extent. The
timber was bought by a company of Irish adven-
turers, who paid at the rate of sixpence a tree
for such as would now have been valued at five
guineas. After having felled the whole of the
forest, the purchasers became bankrupt, and dispersed; the overseer of the workmen was hanged
at Inverary for assassinating one of his men ; the
laird never received the purchase-money of his
timber, and a considerable number of the trees
were left upon the spot where they fell, or by the
shores of Loch Awe, whither they had been carried
for conveyance, and gradually consumed by the
action of the weather. The mosses where the
ancient forests formerly stood are filled with the
short stumps of trees still standing where they
grew. Age has rendered them almost rotten to
the core, and the rains and decay have cleared
them of the soil; yet their wasted stumps and
the fangs of their roots retain their original shape.
Abundance of similar remains are to be seen in
other parts of the Highlands, sometimes inter-
spersed with living and flourishing trees, but surrounded on all sides by the shattered stumps, fallen
trunks, and blasted limbs of a departed forest.*
    A like fate has overtaken the forest of Glen-

* J. H. Allan's " Last Deer of Beann Doran."

more, once famous for the size and age of its
timber, whose magnificent Pines clothed one of
the romantic glens between the Cairngorum range
and the river Spey. This noble forest was purchased of the Duke of Gordon in 1783, and furnished materials for building no less than forty-
one sail of ships, including a frigate of one thousand and fifty tons. A specimen of timber from
one of these trees, preserved in Gordon Castle, is
six feet two inches long, and five feet five inches
broad, with the texture of the finest Red-wood
Pine, and shewing annual growths to the number
of two hundred and thirty-five. The spot was
visited about twenty-five years since by Mr. Selby,
who thus describes its appearance :—" Scattered
trees, some of which were in a scathed or dying
state, of huge dimensions, picturesque in appearance from their knotty trunks, tortuous branches,
and wide-spreading heads, were seen in different
directions, at unequal and frequently at considerable distances from each other : the solitary and
mournful-looking relics of the departed glories
of this once well-clad woodland scene, and which
had only escaped the axe from their previous
decay, or the comparative worthlessness of their
knotty trunks; while the surface of the ground in
almost every direction was littered and bristling
with the decaying tops and loppings of the felled
trees, among which, mosses of various species were
growing with a luxuriance we never saw equalled—
nourished, it would appear, and encouraged by
the partial stoppage and stagnation of the surface-
water thus impeded in its course, and threatening
to convert a large proportion of the surface that
had once been forest into a peat moss." Sir

T. D. Lauder, describing the same scene, says :—
" Many gigantic skeletons of trees, above twenty
feet in circumference, but which had been so far
decayed, at the time the forest was felled, as to be
unfit for timber, had been left standing, most of
them in prominent situations, their bark in a great
measure gone,—many of them without leaves, and
catching a pale unearthly-looking light upon their
grey trunks and bare arms, which were stretched
forth towards the sky like those of wizards, as if
in the act of conjuring up the storm which was
gathering in the bosom of the mountains, and
which was about to burst forth at their call."

Tradition favours the Pine's being considered a
native Forest Tree of England as well as of Scotland.   Gerard says : " I have seene these trees
growing in Cheshire, Staffordshire, and Lancashire,
where they grew in great plentie, as it is reported,
before Noah's floud, but, then being overflowed
and overwhelmed, have been since in the mosses
and waterie moorish grounds, very sound and
fresh untill this day; and so full of a resinous
substance, that they burne like a torch or linke,
and the inhabitants of those countries do call it
Firre wood and *Fire* woode unto this day."

Logs of Pine-wood intermixed with brick have
also been found embedded in the soil, and serving
as the foundation of an ancient Roman road.
Pine-woods are scarcely to be found in England
of so romantic a character as the Highland Forests;
but some of the wilds of Wiltshire and other
English counties are covered with these trees,
self-sown and unpruned, and presenting on a less
grand scale many of the features described as
characteristic of the Scotch forests.

That the Scotch Fir was formerly very abundant in Ireland is proved by the vast quantities of timber still found in many of the extensive bogs for which that country is noted. In the counties of Down, Fermanagh, Donegal, Sligo, Antrim, &c., peat-cutters frequently arrive at layers of these trees in different states of preservation; some are much decayed, others are perfectly sound, and measure as much as seventy feet in length. The depth at which they lie beneath the surface varies from eight to fifteen feet. In some instances they all lie with the top towards the north, the base of the trunks and the upper parts of the stumps, which still remain fixed in the peat, bearing evident marks of fire. Some had attained a large size before they fell, measuring from eight to twelve feet in circumference; in other instances, where the trunk has decayed, the stumps are found imbedded in the peats till quite sound, the roots averaging more than a foot in diameter, and occupying a space varying from thirty to ninety feet in circumference, but never descending to any considerable depth. A single stump frequently furnishes from sixty to seventy bushels of chips. Trunks of Oak are often found lying in the gravel beneath the peat, but Fir has never been noticed in such situations. These trees are invariably rooted in the peat, but at various depths, evidently proving that their growth did not commence until the bog was actually in the course of formation, and that they succeeded each other as in the Highland forests. Instances, indeed, occur in which a large stump is fixed in the peat immediately over another; more rarely a prostrate trunk is

found at such a distance beneath the roots of
another that more than a century must have

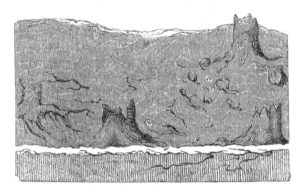

SECTION OF A BOG CONTAINING FIR-STUMPS.

elapsed between the destruction of the first and
the growth of its successor.

In the bog districts, the wood obtained from
these sources forms the principal fuel. It makes
a brilliant and fragrant fire, a property of no
little value in the cheerless districts where it
abounds; though one is by no means disposed,
on that account, to forgive the bogs for having
swallowed up the noble forests, the place of which
they have usurped. This wood is also much used
as a building material, especially when it is likely
to be exposed to wet, its long seasoning having
rendered it indestructible by damp. For the same
reason it is preferred to any other wood for making washing-tubs, axles of mill-wheels, &c.

The range of the Pine is very extensive; it
is to be met with throughout the greater part

II.   A A

of Europe, from the Mediterranean to Norway,
varying in elevation from seven hundred to nearly
four thousand feet, in favourable situations attaining a height of a hundred feet or more, with a
trunk upwards of four feet in diameter, and
dwindling, as it ascends the mountains, to a mere
bush. A variety is said to grow at Nootka Sound
in North America, and it is found also in Siberia,
Kamschatka, Caucasus, and Japan. There are
immense forests of it on the table-lands of Russia,
and on most of the mountain ranges of Europe, as
far south as the Pyrenees. The seeds are sometimes carried by the wind from these latter situations
to marshy places and peat-bogs; but here, though
the seeds germinate, the trees are always stunted
in growth, and soon sicken and die. The finest
specimens grow in a dry soil, and it has been remarked that in native forests the roots run along
the surface, and even rise above it; and the tree
seems to derive a great part of its nourishment
from the black vegetable mould formed by the
decay of its own leaves. The trunk is generally
straight, and covered with a scaly bark of a reddish
hue. The leaves grow in pairs, sheathed at the
base, from two to three inches in length on young
trees, but in old trees they are much shorter. They
are convex on one side, and nearly flat on the
other, so that when pressed together, they form a
cylinder; the edges are minutely notched, and the
colour is a light bluish green, especially beneath,
or on the convex surface. They remain attached
to the tree four years, and, long before this, exchange the glaucous hue for a dark green. The
flowers appear in May and June, the barren ones
arranged in whorls around the extremities of the

last year's shoots, and producing pollen in great
abundance.   The fertile catkins grow most frequently in pairs at the summit of the new shoots,
and gradually assume the form of cones, which are
not ripe until eighteen months old.   They are
stalked, brown, rugged, and more or less tapering

SCOTCH FIR.

to a point.   In the autumn of the second year
they begin to open at the extremity, and shed the
seeds, which are situated in pairs at the base of
each scale : they are small, and furnished each
with a long membranous wing.

There are two principal varieties of the Scotch Fir: in one, the trunk is red and nearly smooth, the branches form a pyramidal head, and the cones are abundant, tapering almost to a point; in the other, the trunk is rugged and yellowish brown, the branches take a horizontal direction or bend downwards, the cones are less numerous and not so much pointed, and the leaves are shorter, of a much lighter, and decidedly glaucous, hue. The timber of the former variety is white, soft, and of little value; that of the latter is red, firm, resinous, and durable.

In the natural forests of Pine, the young plants, being the produce of different years, and consequently of various sizes, the stronger gradually destroy the weaker, until the wood is reduced to the distances at which the trees can ultimately stand, whilst the lateral branches gradually decay and fall off, so that thinning and pruning are quite unnecessary. In short, a natural or self-sown forest of Pines is left entirely to nature. Nature sows the seed, rears the tree, prunes and thins the wood; and the hand of man is applied only to cut it down when fit for timber. In planted woods, the Pines are commonly of the same age and size; and then it is absolutely necessary to thin them, as their tops rise equal, and form a surface parallel to that of the ground on which they stand; therefore, without relief by thinning, the whole are, to a certain extent, injured.

Gilpin considers the Scotch Fir, when in perfection, a very picturesque tree, and admires both the colour of the leaf and its mode of growth. "Its ramification," he says, "is irregular and beautiful, and not unlike that of the Stone

Pine, which it resembles also in the easy sweep of its stem, and likewise the colour of its bark, which is commonly, as it attains age, of a rich reddish brown. The Scotch Fir, indeed, in its stripling state, is less an object of beauty. Its pointed and spiry shoots, during the first years of its growth, are formal; and yet I have sometimes seen a good contrast produced between its spiry points and the round-headed Oaks and Elms in its neighbourhood. When I speak, however, of the Scotch Fir as a beautiful individual, I conceive it when it has outgrown all the improprieties of its youth, when it has completed its full age, and when, like Ezekiel's Cedar, it has formed its head among the thick branches. This character of the Scotch Fir is true only when it grows singly, or in small groups; for when the trees grow in compact bodies, the heads are drawn up without forming lateral branches, the stems becoming mere poles, with heads of tufted foliage; consequently, the trees which furnish the most valuable timber, possess little picturesque beauty."

" Few persons," says Mr. Forsyth, " can form any correct idea of the true character of the Mountain Pine from the samples generally seen in English screen plantations, where this noble tree is made the nurse and drudge of something else for which the soil and climate are better adapted, the muddy swamp and arid barren steep being alike unfavourable to its perfection. Nor need we look to plantations in England and Wales for better examples, for though the soil and climate may be suitable, the systematic butchering of the young trees by the scientific pruner renders the forest only a host of stalwart cripples. Common sense and ob-

servation cry aloud that trees abounding in resin
and gum must be injured by pruning, which is
only another name for wounding."

The timber of the Scotch Fir, especially the
horizontal variety described above, (which is generally considered to be the true Highland Pine,)
is similar in every respect to the best Baltic Pine,
and is highly prized. The best is obtained from
trees the age of which averages about a hundred
and twenty years, and which, from their growing
in a cold climate, have matured their timber
slowly. The earlier the age at which the side
branches die and drop off, the clearer is the wood
of knots, and proportionally the more valuable.
When fully matured, it is of a red hue, and is
considered scarcely less valuable than the Oak,
instances being on record where timbers of Pine
in the roofs of old buildings have, after the lapse
of several centuries, been found perfectly sound.
It is light, stiff, and strong, easily worked, and
freer from knots than that of any other kind of
Fir,—qualities which render it admirably adapted
for all kinds of house carpentry. Its size, length,
and straightness of trunk, fit it also for the main
timbers of buildings, such as rafters, joists, &c.,
which are almost universally made of it. In naval
architecture it is very extensively used, and the
best masts are considered to be those made of the
Pine imported from the Baltic. In Russia, many
of the roads are formed of the trunks of the Pine,
trees being selected which are from six to twelve
inches in diameter at their largest end. The ground
being marked out for the road, the trunks are laid
down side by side, the thick end of the one alternately with the narrow end of the other, and the

branches being left at the summit to form a sort
of hedge on each side of the road, which is very
useful as a guide to travellers when the ground is
covered with snow. The interstices are then filled
in with earth, and the road is finished. In Lapland
and Northern Russia, the outer bark, like that of
the Birch, is frequently used by the natives for
covering their huts, or as a substitute for cork, to
float the nets of the fishermen. The inner bark
is made into ropes, and sometimes woven into
mats, like those made from the Lime-tree. In Norway where it is the custom to kiln-dry oats to such
a degree, that both the grain and the husks are
made into a meal almost as fine as wheaten flour,
in seasons of scarcity, the dried inner bark of the
Pine is ground with the oats, and made into thin
cakes, which, when baked upon a girdle, are said
to be not unpalatable.

From the growing tree turpentine may be procured by stripping off a piece of bark from the
trunk in spring, when the sap is in motion, and
the resinous juice that exudes is received in a notch
or hollow cut in the tree; this juice, as it accu-
mulates, is ladled out into a basket, and the liquid
that passes through is the common turpentine.
The thick matter which remains is distilled with
water, and produces spirit of turpentine, leaving
the common yellow resin of the shops. But the
greatest quantity of turpentine used in this
country is imported from America, where it is
obtained from the Carolina Pine.

Tar is obtained from the wood of the Pine after
it has been felled; Dr. Clarke thus describes the
method of procuring it :—" The inlets of the Gulf
of Bothnia everywhere appeared of the grandest

character, surrounded by noble forests, whose tall trees, flourishing luxuriantly, covered the soil quite down to the water's edge. From the most southern parts of Westro-Bothnia to the northern extremity of the gulf, the inhabitants are occupied in the manufacture of tar, proofs of which are visible along the whole extent of the coast. The process by which the tar is obtained is very simple; and, as we after witnessed it, we shall now describe it from a tar-work we halted to inspect upon the spot. The situation most favourable to the process is in a forest near to a marsh or bog, because the roots of the Scotch Pine, from which tar is principally extracted, are always most productive in such places. A conical cavity is there made in the ground (generally in the side of a bank or sloping hill); and the roots, together with logs and billets of the wood, being neatly trussed in a stack of the same conical shape, are let into this cavity. The whole is then covered with turf, to prevent the volatile parts from being dissipated, which, by means of a heavy wooden mallet, and a wooden stamper, worked separately by two men, is beaten down, and rendered as firm as possible above the wood. The stack of billets is then kindled, and a slow combustion of the kiln takes place, as in making charcoal. During this combustion, the tar exudes; and, a cast-iron pan being fixed at the bottom of the funnel, with a spout that projects through the side of the bank, barrels are placed beneath this spout to collect the fluid as it comes away. As fast as these barrels are filled, they are bunged, and are then ready for immediate exportation. From this description it will be evident that the mode of obtaining tar is

by a kind of distillation *per descensum* (downwards),
the turpentine, melted by fire, mixing with the
sap and juices of the Pine, while the wood itself,
becoming charred, is converted into charcoal."
Dr. Clarke, after stating that tar was made by the
Greeks more than two thousand years ago, remarks that " there is not the smallest difference
between a tar-work in the forests of Westro-
Bothnia and those of Ancient Greece. The Greeks
made stacks of Pine, and, having covered them
with turf, suffered them to burn in the same
smothered manner; while the tar, melting, fell to
the bottom of the stack, and ran out by a small
channel cut for the purpose."

The country people of Scotland obtain tar by a
method similar in principle to that above described, although differing slightly in the details.
They hew the wood into billets, put these into a
pit dug in the earth, and ignite them: the top is
covered with rude tiles; and the tar, as it leaves
the wood, flows out through a small orifice at the
bottom of the pit.   When pitch is to be made,
the tar is put into large copper vessels, and is then
suffered to boil for some time, the volatile part
flies off, and what remains, when cold, hardens
and becomes pitch.

In seasons of scarcity, the bark of the Pine is
converted by the Swedish peasants into bread.

" In the character of the Swedish peasant many
traits present themselves well worthy of imitation in the other ranks of society.   Placed in a
part of the world where the influence of winter
is felt for more than half the year, and where the
general barrenness of the soil must necessarily
subject him to great privations, he is, notwith-

standing, cheerful and contented. In the northern parts, where the early approach of the frost, even in the midst of summer, sometimes cuts off the whole of his scanty crop, and deprives him of his winter provision, he finds bread even in the heart of the forest; and with the bitter bark of the Pine, beaten till it is reduced to a fine pulp, he continues to support existence, living by means of this unpalatable food where others would die. Fortunately it is only in years of great scarcity that he is compelled to have recourse to these means; nor did I, during my travels in the north, ever meet with this *barke bröd*, or bark bread, used as food by the poorer classes. Hard as his fare is at all times, the Swedish peasant exhibits no signs of discontent; and if his countenance do not pourtray a great flow of spirits or hilarity of manner, it shews him to be what he really is, humble, serious, devout, and happy. Give him but the smallest trifle, he receives it with thankfulness, and you are doubly repaid by the grateful and contented manner in which it is accepted."*

* Brook's "Sweden,"

THE PINASTER.

PINUS PINASTER.

The Cluster Pine, or Pinaster.

THE Pinaster is one of the most extensively
planted, in this country, of all the foreign Pines.
In its younger stage it is a pyramidal bushy
tree, well marked by its erect growth and regular
whorls of ascending branches from a foot to a foot
and a half apart, by its tufts of long deep-green
leaves, and by its clusters of large cones, which
are perfected on very young trees. From the
star-like arrangement of these cones it derives its
name of Pin-aster,—Star Pine. The clusters are
situated beneath the whorls or tiers of branches,
and contain from four to a dozen cones; but it is
far from uncommon to see as many as twenty or
thirty in a mass, the lowermost being forced, by
the pressure of those above, to point downwards.
They often remain attached to the tree many
years after they have attained maturity, and indeed may sometimes be seen, covered with grey
lichen, adhering to the main stem, on which, while
it was a mere twig, they were produced a dozen
years before. For the first five and twenty years
of its growth, the age of the Pinaster may be
discovered with tolerable accuracy from observing
the number of tiers formed by its branches, each
interval between two tiers being the result of a
year's growth. As the tree grows older, the lower

limbs die off, and the trunk becomes covered with a purplish bark, marked with numerous deep fissures, and, in exposed situations, often invested

CONES OF PINASTER.

with the grey lichen alluded to above. The bark itself is of a soft pithy texture and readily splits into plates about two inches wide, and from four to six in length, having an even surface on both sides. From the number of these plates also the age of a tree may be nearly computed, for unless any of the outer scales have peeled off, which sometimes happens, the age of the tree corresponds with the number of annual deposits. The trunks of old trees generally incline a little on one side, this effect having been produced by the weight of the foliage &c., while they were young. The lower part of the trunk is entirely bare of branches, but higher up there usually project the stumps of numerous dead branches of unequal lengths and diameters, and the head bears a close resemblance to that described below as characterizing the Stone Pine, except that it does not spread so widely. The roots are few in number, but unusually stout, and instead of extending themselves laterally, as is the case with most of the Fir tribe, they descend almost perpendicularly. Consequently the Pinaster does not flourish on a thin soil, but delights in a dry and sandy situation.

The Pinaster inhabits a wide range of country, being found in the south of Europe, the north of Africa, and the west of Asia; it is also said to grow on the Himalayan mountains. Great use has been made of this tree, in the south of France, in fixing the shifting surface of the sand-hills, and even in turning the waste land which they occupied to profitable account. In the neighbourhood of the Gulf of Gascony alone, there were, in 1789, no less than three hundred square miles rendered worse than useless by innumerable naked sand-hills, which were constantly altering their position, and on the occurrence of storms, having their

surface blown inland to the great detriment of
the cultivated lands.   The remedy proposed by
M. Bremontier was to erect a fence of hurdles so
as to front the prevailing wind, and to sow within
this a belt of Pinaster seeds mixed with those
of the Yellow Broom.   At a short distance within
this was sown a second and a third belt, till the
whole was covered.   The ground was then, as it
were, roughly thatched with hundreds of trees,
reeds, or sea-weed.   Thus protected, the seeds
sprung up, the Broom at first outstripping its companion and affording it shelter.   In the course of
seven or eight years, it was found that the Pinaster
began to choke its foster-nurse, which quietly
submitted and gave up its decaying leaves and
twigs to the fertilization of the soil.

In about ten or twelve years the plantations
were thinned, the branches being applied to the
sheltering of ground not hitherto enclosed, and
the trunks being burned to make tar.   When about
twenty or thirty years of age, the trees are fit for
producing resin ; and when exhausted for this purpose, they are cut down to make room for their self-
sown progeny.   In this way many thousands of acres
have been reclaimed and converted into plantations, which afford occupation to the inhabitants
of the surrounding districts, who gain their liveli-
hood by the manufacture of resin and tar.   From
its power of resisting the sea-blast, the Pinaster is
sometimes called the Sea Pine (*Pinus maritima*).
I am not aware that its valuable property of
binding sand-hills has been tested in this country,
but in the west of England it is frequently planted
on the sea side of plantations composed of other
trees, and proves an effectual shelter, never shew-

ing the least tendency to bend before the prevailing wind, and never having its outer branches blighted.

The common resin of commerce is extracted from the Pinaster while it is in a growing state. In summer, trees are selected which have a trunk about four feet in diameter, and longitudinal cuts are made through the bark, about six inches wide and a foot long, with a cavity at the base. Into this the resin flows from between the bark and the wood, and is scooped out occasionally with a ladle. It is found necessary to lengthen the cut very frequently, as the resin does not flow freely from an old wound. In a few years the tops of the grooves are too high to be reached by a man standing on the ground; the operator therefore climbs the tree by the help of a notched pole, and when the trees have ceased to produce resin, they are cut down to be manufactured into tar. The resin is melted in caldrons, and strained through straw to free it from impurities; it is then stored away in barrels and is fit for the market.

To make the best lamp-black, the straw through which resin has been strained is put into a stove and kindled: the smoke passes through a chimney into a chamber which has an opening in the roof; over the opening is placed a flannel bag, supported by wooden rods in the form of a pyramid. The soot is deposited either on the walls of the chamber or on the flannel bag,—the flannel acting as a filter to the lighter part of the smoke, by retaining the soot and allowing the heated air to escape. The soot is detached from the flannel bag by striking the outside smartly with a stick; and,

II. B B

the door of the chamber being opened, the lamp-
black is swept out and packed in small barrels.
Tar is sometimes substituted for resinous straw,

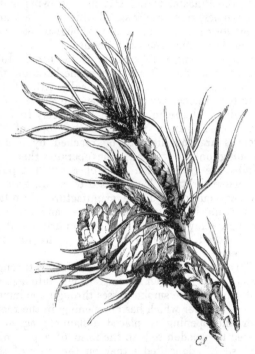

PINUS LEMONIANA.

and lamp-black is sometimes obtained by burning
resin in a kind of lamp furnished with a chimney,
which is surrounded by a flannel, and which retains the soot. It was from this mode of obtain-

ing lamp-black that that substance derived its name.

The Pinaster also produces tar, pitch, and oil of turpentine, but not of a fine quality.

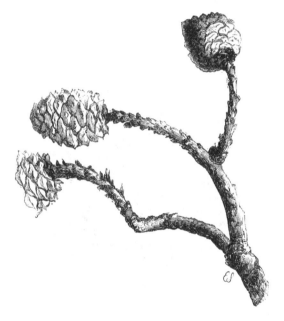

DEAD BRANCH OF PINUS LEMONIANA.

A singular variety of Pinaster was noticed by Sir Charles Lemon at Carclew in Cornwall, which has been named *Pinus Lemoniana.* The peculiarity of this tree is, that it bears at the extremity of every branch a solitary cone instead of a new shoot, the side shoots rising from beneath the base of the cone. Hence the tree has a singular zigzag appearance in its young stage, and, when old, is more bushy than the common Pinaster.

THE STONE PINE.

PINUS PINEA.

THIS is the Pine of Italy and the Tyrol, much
prized for its nuts even in Pliny's time, who says
that it bore at the same time ripening fruit, fruit
destined to be matured the next year, as well as
in that which followed. He adds that ripe ones
might be gathered at all seasons. In its native
country it is described as a noble tree with a
towering stem, often exceeding a hundred feet in
height, and the Latin poets often celebrate it as
" the vast Pine." It throws up a naked tapering
stem, and bears at its head an extended table-like
mass of branches, laden with a peculiarly rich
green foliage. Though introduced into Britain
so long ago as 1548, and far from uncommon in
collections, it rarely, if ever, assumes its native
character. Its usual form with us is a large dense
bush, leafy to the ground, having no main trunk,
but divided just above the roots into several
crooked branches, which often creep along the
ground to some distance before they begin to ascend. The leaves, which are long, grow in pairs;
they are flat on the inner side and convex on the
outer, and when pressed together form a perfect
cylinder.

Gilpin's description of it, therefore, applies

rather to the Italian form of the tree than to specimens which he has seen in Britain :—" The Stone
Pine promises little in its infancy in point of picturesque beauty. It does not, like most of the
Fir species, give an early indica-
tion of its future form. In its
youth it is dwarfish and round-
headed, with a short stem, and
has rather the shape of a full-
grown bush than of an increasing
tree. As it grows older, it does
not soon lay aside its formal shape.
It is long a bush, though somewhat more irregular, and with a
longer stem; but as it attains
maturity, its picturesque form
increases fast. Its lengthening
stem assumes commonly an easy
sweep. It seldom, indeed, de-
viates much from a straight line,
but that gentle deviation is very
graceful, and above all lines difficult to imitate. If accidentally
either the stem or any of the
larger branches take a larger
sweep than usual, the sweep
seldom fails to be graceful. It
is also among the beauties of the
Stone Pine, that, as the lateral branches decay,
they leave generally stumps, which, standing out
in various parts of the stem, break the continuity
of its lines. The bark is smoother than that of
any other tree of the Pine kind, except the Wey-
mouth; though we do not esteem this among its
picturesque beauties. Its hue, however, which is

warm and reddish, has a good effect; and it obtains a kind of roughness by peeling off in patches. The foliage of the Stone Pine is as beautiful as the stem. Its colour is a deep warm green; and its form, instead of breaking into acute angles, like many of the Pine race, is moulded into a flowing line by an assemblage of small masses. As age comes on, its round clump-head becomes more flat, spreading itself into a canopy, which is a form equally becoming." The cones are larger than those of the Pinaster, of a lighter colour, and more orbicular;* the nuts are three quarters of an inch in length, and furnished with a very short wing. The seeds, after being detached from their strong outer shell, are commonly sold in large quantities all the winter in Florence, Pisa, and other places within reach of the extensive forests of this Pine, under the name of Pinocchi. They are about the same size as the common hazel-nut, only much more oblong, and not very unlike them in taste, except that they have a slight and not disagreeable resinous flavour. Remains of the kernels were found among the domestic stores during the excavations at Pompeii. Sir George Staunton also informs us that they are much prized by the Chinese. In Italy the empty cones, which are highly inflammable, are commonly used for lighting fires.

* See page 325.

THE SPRUCE FIR.

THE SPRUCE FIR.

ABIES EXCELSA.

THE Spruce Fir was known to the ancients by the name of Picea. Pliny describes it as delighting in a lofty and cold situation. He compares its form to that of the Larch, with moderately long branches, or arms spreading from the main trunk close to the root; but the leaves, he says, are scattered, short, rigid, and prickly, and abound in resin. Being a gloomy tree, its branches were used to attach to doors as a sign of a funeral about to take place.* Under the influence of the sun, it sometimes exudes drops of resin. The timber is used for beams, laths, &c. Linnæus, by a strange oversight, considered the Picea of the ancients identical with our Silver Fir, and the Abies of Pliny and other Latin writers he supposed to be our Spruce Fir: but there can be no doubt that he was here in error, the description quoted above being much more applicable to the tree now under consideration.

The Spruce, or Norway Spruce Fir, is a native of the mountainous parts of Europe and Asia, preferring a moist soil and cold climate. It is most frequent in the north, but is found at a

---

* In Sweden and Norway at the present day, when a funeral is about to take place, the road into the churchyard and to the grave is strewn with these green sprigs, the gathering and selling of which is a sort of trade for poor old people about the towns.

great elevation on the Alps, Pyrenees, and other
mountains of central Europe, flourishing in situations which are too cold and wet for the Scotch
Pine.   In Lapland it grows at an elevation of a
thousand feet, in Norway and Sweden at an elevation of from two to three thousand, and among
the alps of Switzerland it attains perfection at a
much greater height.

The usual form of the Spruce Fir is a perfectly
erect pyramidal tree, upwards of a hundred feet
in height, with a solid trunk, which at the
base is from three to six feet in diameter.   In
young trees the lateral branches are arranged in
regular whorls from the very root to within a short
distance of the summit, which is a solitary spear-
like shoot.   They are at first horizontal, ascending
towards the extremities; but as the tree grows
older, the lower branches decay naturally, and are
thrown off, and the upper ones droop and form
a graceful curve; the spray also droops on both
sides of the leading branch, producing a pleasant
feathery appearance.   The leaves are short and
rigid, scattered singly on all sides of the shoots.
The cones are about six inches long, and at the
base two inches in diameter, tapering, and blunt
at the extremity, and, when ripe, hang downwards
from the ends of the branches.

" In a picturesque point of view, the Spruce
Fir is generally esteemed a more beautiful
and elegant tree than the Scotch Fir; and the
reason, I suppose, is, because it often feathers to
the ground, and grows in a more exact and regular shape.   But this is a principal objection to it.
It often wants both form and variety.   We admire
its floating foliage, in which it sometimes exceeds

all other trees ; but it is rather disagreeable to see
a repetition of these feathery strata, beautiful as
they are, reared tier above tier in regular order,

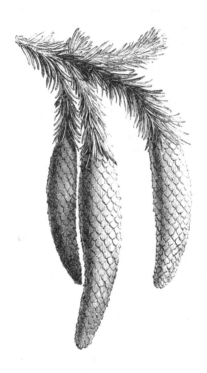

CONES OF SPRUCE FIR.

from the bottom of a tree to the top. Its perpendicular stem, also, which has seldom any
lineal variety, makes the appearance of the tree

still more formal.   It is not always, however, that
the Spruce Fir grows with so much regularity.
Sometimes a lateral branch, here and there, taking the lead beyond the rest, breaks somewhat
through the order commonly observed, and forms
a few chasms, which have a good effect.   When
this is the case, the Spruce Fir ranks among picturesque trees.   Sometimes it has as good an
effect, and in many circumstances a better, when
the contrast appears still stronger—when the tree
is shattered by some accident, has lost many of its
branches, and is scathed and ragged.   A feathery
branch here and there, among broken stumps,
has often an admirable effect, but it must arise
from some particular situation.   In all circumstances, however, the Spruce Fir appears best
either as a single tree, or unmixed with any of its
fellows; for neither it nor any of the spear-
headed race will ever form a beautiful clump
without the assistance of other trees.*

It is, however, only in its native haunts, the
sides of mountain ravines, that the real picturesque beauty of the Spruce Fir can be appre-
ciated; and it is not altogether just of the author
just quoted to measure the excellences of a tree
essentially mountainous by the same rules which
he applies to the humbler inhabitants of the lowlands.   " It is," says Sir T. D. Lauder, " the
great tree of the Alps, and is so mentally associated with the grandeur of Swiss scenery, that
the sight of it never fails to touch chords in our
bosom which awaken the most pleasing recollec-
tions.   What can be more truly sublime than to
behold, opposed to the intensely blue ether, the

* Gilpin.

glazed white summits of Mont Blanc, or the
Jungfrau, rising over the interminable forests of
Spruce Firs, which clothe the bases of the mountains, whilst some such gigantic specimens as those
we have been noticing rise in groups among the
rocks before us, many of them shivered, broken,
and maimed by tempests, their dark forms opposed to all the prismatic hues of some immense
gorgeous glacier, which nourishes in its immense
bosom a mighty river that is doomed to fertilize
and enrich whole kingdoms." It here attains a
height of a hundred and fifty, or even a hundred
and eighty feet, diminishing regularly in size, till
it approaches the boundary of perpetual frost.
The whole of the Hartz Mountains are covered
with it, and it affords both fuel and timber for the
mines and furnaces of that district. It is planted
or sown, and cut down in masses, like our coppice-
woods; and self-sown seedlings supply the vacan-
cies created by every cutting. In France and
Germany, hedges, or rather lines of trees, serving as
boundary-fences, and at the same time as sources
of shelter and shade, are used in the same way as
thorn-hedges are in England. They produce an
enormous quantity of timber for fencing and fuel,
every twenty or thirty years; and every year the
fall of their leaves supplies manure. With us,
however, the Spruce Fir, unless planted in peculiar situations, both as it regards soil and aspect,
does not thrive. It produces abundance of cones
at an early age, but soon dwindles away and dies.

In the country bordering on the Baltic, a Spruce
forest is a very different thing. " This is the
land of Pines,—lofty, erect battalions,—their bark
as smooth as the mast of a ship—their branches

regular as a ladder, varying scarce an inch in girth in fifty feet of growth—for miles interrupted only by a leaning, never by a crooked, tree—with an army of sturdy liliputians clustering round their bases—fifty heads starting up where one yard of light is admitted. What becomes of all the pruning, and trimming, and training—the days of precious labour spent on our own woods? Nature here does all this and immeasurably better for her volunteers, who stand closer, grow faster, and soar higher than the carefully planted and trans-planted children of our soil. Here and there a bare jagged trunk, and a carpet of fresh-hewn boughs beneath, shew where some peasant urchin has indulged in sport which with us would be amenable to the laws, namely, mounted one of these grenadiers of the forest, hewing off every successive bough beneath him, till, perched at a giddy height aloft, he clings to a tapering point which his hand may grasp. The higher he goes, the greater the feat, and the greater the risk to his vagabond neck in descending the noble and mutilated trunk." Sometimes the woods are composed of "mingled trees, the fresh hues of the oak contrasting with the black Pines; and close to us stood a noble Spruce, split from top to base by the lightning of last week's storm, one half resting against a neighbouring stem, the other pale, bleeding, and still erect. Below, lay forty feet of the luxuriant head, with enormous splin-ters, rent in longitudinal lines, while the ground was furrowed in deep angular troughs by the last strength of the fluid. Here and there the sun, shooting across a Silver Birch trunk, like the light across a liquid human eye, or illuminating the red bark of a veteran Scotch Fir with a fiercer glow, or stealing, few and far between, in slender bars of gold along the tender grass."*

The timber of the Spruce Fir has, for an unknown period, been imported into Britain from Norway, chiefly in the form of entire trunks, which are used for scaffolding-poles, spars, oars, and masts for small craft; but partly, also, sawn into planks or deals, known in common as white deal, white Baltic deal, and white Christiania deal, the wood having a red hue only when the tree is grown in certain soils and situations. The poles, spars, and oars are the thinnings of the Norwegian woods; and the deals and planks are made from the larger trees which are left. The slenderest poles are taken from the largest and oldest woods, and are called seedlings; they are always found where the wood is most dense, and very often close by the side of a larger tree. They grow very tall and slender, wholly without branches, except at the summit, and, though often only a few inches in diameter, are of great age.

Nothing can be finer than the Spruce-timber of the Alps, which is so tough, that the natives are actually in the habit of kindling fires about the trees so as to burn them down, to save their own trouble and the edges of the axe.

"In scattered forests," says Loudon, "the snow falling on the trees individually, is retained by their branches, and when these are of great length, often weighs them down and breaks them. We have seldom been more gratified with winter scenery than when passing through a Spruce Fir

* " Letters from the Baltic."

forest in Sweden. We have seen trees of all ages grouped and distributed in innumerable ways,— here weighed down with snow, and there boldly shooting through it their round pyramidal heads. When a sudden thaw takes place in spring, the snow and the branches seem all in motion; some branches, being relieved of their load of snow, are rising up in consequence of their elasticity; and others, from the snow falling on them from branches still higher up the tree, are bending, and perhaps breaking, under the additional weight."

The Spruce Fir, besides furnishing large quantities of valuable timber, produces the substance known by the name of Burgundy pitch. To produce this resin, the collector, in the spring, before the sap begins to ascend, cuts off a long vertical strip of bark from the south side of the tree, as deep as the soft wood, but without wounding it. The sap exudes very slowly from between the bark and the wood, and hardens on exposure to the air. In about three or four months afterwards, the groove is found filled with dry resin, which is then collected and purified by being melted in boiling water.

The uses of deal are too numerous and well known to be noticed here; the bark is used in tanning. In Sweden and Norway the inner bark is made into baskets; and the canoes, which are made of the timber of the large trees, and which are so light that they may be carried on a man's shoulders when the navigation is interrupted by a rapid or cascade, have their planks fastened together by strings made of the roots, so that not a single nail is used in their construction. The long and slender roots are chosen for this purpose; they are rendered flexible by splitting them down the middle, and by boiling them for two or three hours in water, mixed with alkali and sea salt. They are then dried and twisted into cordage, which is used instead of hemp, both for naval and agricultural purposes. From a decoction of the young shoots Spruce beer is made.

Spruce Firs in this country are liable to serious depredations either from squirrels or crossbills, which gnaw off about six inches of the young

GALL OF SPRUCE FIR.

shoots, and let them fall to the ground, sometimes in such quantities as to carpet the soil. It is scarcely decided which of the two animals is the culprit; but the visits of either are much dreaded by foresters. The only insect which injures the Spruce Fir is a kind of aphis, which in the autumn

II.   C C

lays its eggs on the under side of the buds on the side branches. When these begin to burst in spring, the young leaves grow together into a solid mass, composed of a number of cells. Each of these contains an embryo insect, and towards the end of summer opens and suffers the perfect insect to escape. These galls somewhat resemble imperfect cones, bearing a shoot at the summit; the part of the shoot beyond the gall is often distorted in consequence, and sometimes entirely killed. Young trees are the principal sufferers from these attacks. The Laplanders, it is said, eat these galls.

The largest Spruce Fir in Britain is at Studley, and is said to have been planted by Eugene Aram about the middle of the last century. It is a hundred and thirty-two feet high, with a trunk six feet five inches in diameter, clothed with branches from the base to the summit.

THE SILVER FIR.

THE SILVER FIR.

PICEA PECTINATA.

THE Silver Fir approaches in character nearer
to the Norway Spruce than to any other of the
trees yet described. It may, however, be easily
distinguished by the following marks. The leaves,
especially in young trees, are placed singly, but
instead of being inserted on all sides of the
stems, are arranged in two opposite rows, and
thus the surface of the twig is flat instead of convex : the midrib is visible on the under side only,
the upper side having a furrow down its centre.
On each side of the midrib beneath is a white
silvery line, from which the tree derives its name;
and as the point of the leaf is always turned up,
these lines make a conspicuous appearance. The
cones are large and cylindrical, and each scale terminates in a deflexed point. Their position, also,
which is erect, affords also an infallible mark of
distinction, the cones of the Spruce Fir being
pendent. When young they are green, but as
they advance towards maturity they acquire a
rich purplish hue, and when quite ripe are of a
deep brown. They remain upwards of a year on
the tree, appearing in May, and ripening the seed
in the October of the following year. The general outline of the tree, when standing alone, is a
lengthened pyramid. " It has all the regularity
of the Spruce without its floating foliage. There

is a sort of harsh, stiff, unbending formality in the
stem, the branches, and the whole economy of the
tree, which makes it disagreeable.   We rarely see
it, even in its happiest state, assume a picturesque
shape ; assisted it may be in its form when

CONES OF SILVER FIR.

broken and shattered, but it will rarely get rid of
its formality. In old age it stands the best chance
of attaining beauty.   We sometimes see it, under
that circumstance, a noble shattered tree, finely
adorned with ivy, and shooting out a few horizontal branches, on which its meagre foliage and
tufted moss appear to advantage."* This formal
character is to be attributed to the horizontal
direction of its branches from the main stem, and
to the same position of the spray with reference

* Gilpin.

to the branches. The buds, which are situated at the extremities of the shoots, expand in spring, and the young leaves are of a pale green, almost as delicate as the tint of the primrose. I have observed that the Sulphur butterfly, which makes its appearance about the same time with these tender shoots, loves to settle on their under side, either attracted by their flower-like semblance, or taught by instinct to rest where its enemies may be unable to distinguish its yellow wings from the surrounding foliage. If driven from one of these places of retreat, it flies a short distance and alights on another.

The Silver Fir was called by the Romans Abies, a name which, by an error of the early botanists, was given to the Spruce Fir, and hence considerable confusion has arisen. It was much used by the ancients in ship-building, and was considered by Virgil the fairest ornament of the mountains. It is a native of the mountainous parts of central Europe, and of the west and north of Asia, but does not extend so far north as the Spruce or Scotch Fir, nor, from its preferring a milder climate, is it found at so great an elevation as these trees. In dimensions it is one of the most striking of the tribe, rising frequently to the height of a hundred and sixty or even a hundred and eighty feet, with a stem perfectly erect and generally clothed from the base to the summit with regular tiers of horizontal branches, and often measuring as much as six or eight feet in diameter. For many years the bark is smooth and of a greenish grey colour; but as the tree gains age, it becomes rough, with small fissures; and when very old it often throws off the outer part in large flakes, leaving the recent bark of a deep rich brown. Pliny relates that an enormous ship, which was built to transport an obelisk from Egypt to Rome, had for its mast a Silver Fir, measuring twenty-four feet in circumference. It is supposed to have been introduced into England about the beginning of the seventeenth century, and is now very common. Specimens are in existence which have attained a height of a hundred and fifty feet, with a trunk sixteen feet in diameter. " The timber of British growth is found to be of excellent quality, and adapted for almost all purposes to which the wood of the Pine is applicable; it possesses both elasticity and strength; its grain being straight and even, it is not subject to warp or twist, even when sawn out of the green or newly cut log. On the Continent the forests of Silver Fir, besides affording a large supply of naval timber for masts, yards, &c., produce much of the wood used in building; and as it is found to endure a long time when driven as piles under water, it is extensively used for that purpose in Holland and other places."*

The resinous products of the Silver Fir are highly valuable. The substance called Strasburg turpentine (from a large forest of these trees near Strasburg) is collected from small tumours or blisters under the cuticle of the bark. The method of procuring this is thus described by Loudon:—"Every year, about the month of August, the Italian peasants who live near the Alps make a journey into the mountains to collect the turpentine. They carry in their hands cornets of tin, terminating in a sharp point, and a bottle of

* Selby.

the same metal suspended to the girdle round
their waists. Thus accoutred, the peasants climb
to the summits of the loftiest Silver Firs, their
shoes being armed with cramping-irons, like spurs,
which enter into the bark of the trees and thus
support the climber, who also clings to the trunk
of the tree with his knees and one arm, while,
with his other hand, he presses his cornet to the
little tumours which he finds in the bark, to extract the turpentine within them. As soon as a
cornet is filled with the clear turpentine, it is
emptied into the tin bottle which is suspended
from his waist; and when this bottle is full, its
contents are strained into a large leathern bottle
of goatskin. The straining is to free the turpentine from the leaves and bits of bark and moss
which may have fallen into it; and this is the
only preparation that is given to this kind of turpentine, which is kept in the goatskins for sale.
Besides the turpentine collected from the tumours
or blisters, an inferior kind is produced by slightly
wounding the bark of the tree. In rich soils the
trees will yield their sap twice a year, namely, in
spring and August; but, in general, the tumours
are formed only once a year, namely, in spring,
and are full of turpentine in August. The
tumours are sometimes round, and sometimes
oval; but when the latter, their greatest length is
always in a horizontal direction. It is employed,
as well as the essential oil which is distilled from
it, both in medicine and in the arts. It is
the only kind of turpentine, produced by any
kind of Pine or Fir, which is used in the preparation of the clear varnishes, and by artists for
their colours." The bark may be employed for

tanning leather, and is used generally in some parts of Switzerland. In some parts of Europe, the young cones, reduced by boiling to a pulp, and preserved with sugar, are eaten as a sweet-meat.

The Silver Fir is very liable to the attack of an insect belonging to the genus *Eriosoma,* which is not only injurious by disfiguring the individual it infests, but frequently causes the death of the tree by the absorption of its juices. It always attacks the main trunk or the undersides of the branches; appearing in patches, and covered with a white cottony substance. Not only are young trees killed by these destructive insects, but full-grown trees measuring eleven or twelve feet in girth. The only remedy which has been found effectual is to rub the insects to death with a brush or coarse canvas cloth in spring. It does not appear to extend its ravages to any others of the Fir tribe, nor has it been known for more than thirty or forty years; but it is said to be greatly on the increase.

THE LARCH.

THE LARCH.

LARIX EUROPÆA.

THE Larch may best be distinguished from the rest of the Fir-trees at the season when most other trees throw off their distinctive character. In winter its lofty undivided stem, pyramidal form, and tiers of drooping branches still bearing the cones formed during the preceding summer, decisively attest its relationship with the Firs; and the absence of leaves at once distinguishes it from any other of that tribe with which we are familiar. There is, however, no difficulty in detecting it, no matter what may be its associates, when in full foliage. A favourable specimen of the Larch may be described as an erect tree, of a pyramidal form, clothed with long slender branches from its pointed summit almost to

TWIG OF LARCH.

the ground, the lower ones being more or less pendulous, as also is the spray throughout. The leaves are bright green, growing in tufts, of a soft texture, spreading, and slightly recurved. The cones, which are small, are numerous, and arranged along the twigs in rows more or less regular. In their young state they vary in colour from greenish white to bright red, and when ripe are brown, being formed of overlapping scales, which are not united into a compact woody mass, but are detached at the edges.

Though it possesses little claim to picturesque beauty,—at least, in its British garb,—" it must be acknowledged," says Wordsworth, " that the Larch, till it has outgrown the size of a shrub, shews, when looked at singly, some elegance in form and appearance, especially in spring, decorated as it then is by the pink tassels of its blossoms; but, as a tree, it is less than any other pleasing. Its branches (for boughs it has none) have no variety in the youth of the tree, and little dignity even when it attains its full growth. Leaves it cannot be said to have ; and consequently it affords neither shade nor shelter. In spring the Larch becomes green long before the native trees ; and its green is so peculiar and vivid, that finding nothing to harmonize with it wherever it comes forth, a disagreeable speck is produced. In summer, when all the other trees are in their pride, it is of a dingy, lifeless hue; in autumn of a spiritless unvaried yellow ; and in winter, it is still more lamentably distinguished from every other deciduous tree of the forest; for they seem only to sleep, but the Larch appears absolutely dead."

In its native haunts, however, the Alps, and other mountains of central Europe, it occupies an important position, growing abundantly in the chasms and ravines, especially on the north sides of the mountains, and striving to impart to these lonely regions the solemn character with which the Silver Fir clothes the south. It here rises to the height of eighty or a hundred feet, with a trunk from three to four feet in diameter. As it grows naturally on the Appennines it was known to the Romans, and is repeatedly mentioned by Pliny as a lofty deciduous tree, valuable for the strength and durability of its timber, but worthless as fuel, its wood being not convertible into charcoal, and as uninflammable as a stone. "We read," says Evelyn, "of beams of no less than an hundred and twenty feet in length, made out of this goodly tree, which is of so strange a composition that it will hardly burn. Yet the coals thereof were held far better than any other for the melting of iron, and the locksmith. There is abundance of this Larch timber in the buildings at Venice, especially about the palaces in Piazza San Marco. Nor did they only use it in houses, but in naval architecture also. The ship mentioned by Witsen to have been found not long since in the Numidian Sea, twelve fathoms under water, was chiefly built of this timber and Cypress, both reduced to that induration and hardness as greatly to resist the fire and the sharpest tool; nor was anything perished of it, though it had lain above a thousand and four hundred years submerged. Tiberius, we find, built that famous bridge to his Naumachia with this wood; and it seems to excel for beams, doors,

windows, and masts of ships : it resists the worm.
Being driven into the ground, it is almost petri-
fied, and will support an incredible weight;
which, and for its property of long resisting fire,
makes Vitruvius wish they had greater plenty of
it at Rome to make joists of; for that, being attempted with fire, it is long in taking hold, growing only black without. It makes everlasting
spouts and pent-houses, which need neither pitch
nor painting to preserve them; also excellent
pales, posts, rails, pediments, and props for vines;
to these add the palettes on which our painters
blend their colours. Before the use of canvas and
bed-tick, it formed the tables on which the great
Raphael and the famous artists of the last age
eternized their skill."

In Evelyn's time the value of British-grown
Larch had not been tested; for though he says
" We grow it of seeds," it is clear from what he
afterwards says that it was of uncommon occurrence. " That it flourishes with us, a tree of good
stature, not long since to be seen about Chelms-
ford, in Essex, sufficiently reproaches our not
cultivating so useful a material for many purposes, when lasting and substantial timber is
required."

About the middle of the last century some
trees, planted by the Duke of Athol, were cut
down, and the timber was found to be superior
to that of any other of the Fir-tribe. A further
acquaintance with the tree confirmed this opinion,
and Loudon tells us that it has been more extensively planted in Britain, particularly since
the commencement of the present century, than
any other timber-tree whatever, not even except-

ing the Oak. John, Duke of Athol, successor of the duke mentioned above, planted between the years 1764 and 1826, the enormous number of 14,096,719 Larches, occupying a space of 8,604 Scotch acres, or 10,324 imperial acres. " There is no name that stands so high, and so deservedly high, in the list of successful planters, as that of the late John, Duke of Athol. His grace planted, in the last years of his life, 6,500 Scotch acres of mountain ground *solely with Larch*, which, in the course of seventy-two years from the time of planting, will be a forest of timber fit for the building of the largest class of ships in his Majesty's navy. Before it is cut down for this purpose, it will have been thinned out to about four hundred trees per acre. Each tree will contain at the least fifty cubic feet, or one load of timber, which, at the low price of one shilling per cubic foot (only half of its present value), will give a thousand pounds per acre; or, in all, a sum of 6,500,000*l.* sterling. Besides this, there will have been a return of seven pounds per acre from the thinnings, after deducting all expense of thinning, and the original outlay of planting. Further still, the land on which the Larch is planted is not worth above from ninepence to one shilling per acre. After the thinnings of the first thirty years, the Larch will make it worth at least ten shillings an acre, by the improvement of the pasturage, upon which cattle can be kept both summer and winter."*

Mr. Gorrie, who admits that the above statement of the probable value of the Larch timber is over-estimated, remarks that Larch is by far the

* "Transactions of the Highland Society."

best improver of heath or moor-pasturage known
in this country. If planted thick, it will in a few
years choke the heath and coarser grasses; and
these plants will be succeeded by finer grasses,
with a foliage possessing a softness and luxuriance
never acquired in open situations. The Larch
ripens its seeds freely in Britain, and is now
raised by the Scotch nurserymen in larger quantities than any other timber-tree.

Larch timber is said to be superior to foreign
Fir in the following respects: it is clearer of
knots, more durable, the dead branches even being
never found to be rotten; it is much less liable
to shrink or split; it may be seasoned in a much
shorter time; it is more tough; it is of a better
colour, and susceptible of a polish superior to
that of the finest mahogany, and more durable,
bearing exposure to changes of climate and
moisture for many years without undergoing any
change.

From possessing these properties it is considered
by good judges to be better adapted for naval
architecture than any other timber. It becomes
harder and more durable by age in a ship. It
holds iron as firmly as Oak, but, unlike Oak, it
does not corrode iron. It does not shrink ; it possesses the valuable property of resisting damp.
It catches fire with difficulty, and it does not
splinter when struck by a cannon-ball. These
qualities have been tested in the case of the
Athol, a twenty-gun frigate, which was launched
in 1820, the keel, masts, and yards of which were
made wholly of Larch.

The timber is found to be equally well suited for
house carpentry, joining, &c.; and for hop-poles,

vine-props, and rails for fencing, it is preferred to any other wood, bearing exposure to all weathers without shewing any symptom of decay in the course of many years. For the same reason it is in great demand for the sleepers of railways. The bark possesses tanning properties to a considerable extent, but being in this respect far inferior to Oak, it will not pay the expenses of peeling and carriage.

From the trunk of the full-grown Larch is procured the substance known by the name of Venice turpentine. This is a liquid resin found in large cavities, which measure sometimes several inches across, and are situated in the solid wood, five or six inches from the heart of the tree. In order to obtain it, holes are pierced with augers, and into them are inserted wooden tubes, through which the turpentine flows into little buckets suspended at the other end to receive it. The season for collecting it lasts from May to October. It is perfectly clear, and needs no further preparation than straining through a coarse hair cloth to free it from impurities. It derived its name from the City of Venice, from which it was formerly exclusively exported. It is used in medicine, and for making several kinds of varnish.

A manna is also produced from the shoots of the young Larches, which resembles that of the Ash; it is called Manna of Briançon, from the name of the place where it is collected.

CHELSEA CEDAR.

THE CEDAR OF LEBANON.

CEDRUS LIBANI.

MANY years ago a Frenchman, who was travelling in the Holy Land, found a little seedling
among the Cedars of Lebanon, which he longed
to bring away as a memorial of his travels. He
took it up tenderly, with all the earth about its
little roots, and, for want of a better flower-pot,
planted it carefully in his hat, and there he kept
it and tended it.

The voyage home was rough and tempestuous,
and so much longer than usual, that the supply of
fresh water in the ship fell short, and they were
obliged to measure it out most carefully to each
person. The captain was allowed two glasses a-
day, the sailors, who had the work of the ship on
their hands, one glass each, and the poor passengers but half a glass. In such a scarcity you may
suppose the poor Cedar had no allowance at all.
But our friend the traveller felt for it as his
child, and each day shared with it his small half
glass of precious water ; and so it was, that when
the vessel arrived at the port, the traveller had
drunk so little water that he was almost dying,
and the young Cedar so much that, behold, it was
a noble and fresh little tree, six inches high !

At the Custom-house the officers, who are always suspicious of smuggling, wished to empty
the hat, for they would not believe but that some-

thing more valuable in their eyes lay hid beneath
the moist mould. They thought of lace, or of
diamonds, and began to thrust their fingers into
the soil. But our poor traveller implored them
so earnestly to spare his tree, and talked to them
so eloquently of all that we read in the Bible of
the Cedar of Lebanon, telling them of David's
house and Solomon's temple, that the men's hearts
were softened, and they suffered the young Cedar
to remain undisturbed in its strange dwelling.
From thence it was carried to Paris, and planted
most carefully in the Jardin des Plantes. A large
tile was set against it as a protection and a shade,
and its name was written in Latin and stuck in
front, to tell all the world that it was something
new and precious. The soil was good and the
tree grew, grew till it no longer needed the shelter
of the tile, nor the dignified protection of the
Latin inscription; grew till it was taller than its
kind protector the traveller; grew till it could
give shelter to a nurse and her child, tired of
walking about in the pleasant gardens, and glad of
the coolness of the thick dark branches. The
Cedar grew larger and larger, and became the
noblest tree there. All the birds of the garden could have assembled in its branches. All the
lions and tigers, and apes and bears, and panthers
and elephants of the great menagerie close at
hand, could have lain at ease under its shade. It
became the tree of all the trees in the wide garden that the people loved the best; there, each
Thursday, when the gardens were open to all the
city, the blind people from their asylum used to
ask to be brought under the Cedar; there they
would stand together and measure its great trunk,

and guess how large and wide must be its branches.
It was a pleasure to see them listening to the
sweet song of the birds overhead, and breathing
in its fragrant eastern perfume. There was once
a prison at the end of these gardens, a dark and
dismal and terrible place, where the unfortunate
and the guilty were all mixed together in one
wretched confusion. The building was a lofty
one, divided into many stories, and, by the time
you reached the top, you were exhausted and
breathless. The cells were as dreary and com-
fortless there as in the more accessible ones below,
and yet those who could procure a little money by
any means, gladly paid it to be allowed to rent
one of these topmost cells. What was it that
made them value this weary height? It was, that
beyond that forest of chimneys and desert plain
of slates, they could see the Cedar of Lebanon!
His cheeks pressed against the rusty bars, the
poor debtor would pass hours looking upon the
Cedar. It was the prisoner's garden, and he
would console himself in the weariness of a long,
rainy, sunless day, in thinking, "The Cedar will
look greener to-morrow." Every friend and
visitor was shewn the Cedar, and each felt it a
comfort in the midst of so much wretchedness to
see it. They were as proud of the Cedar
in this prison as if they had planted it.
Who will not grieve for the fate of the
Cedar of Lebanon? It had grown and flourished
for a hundred years, for Cedars do not need centuries, like the Oak, to attain their highest growth,
when, just as its hundredth year was attained, the
noble, the beautiful tree was cut down to make
room for a railway! This was done just twelve

years ago; and now the hissing steam-engine
passes over its withered roots. Such things, it
seems, must be; and we must not too much
grieve or complain at any of the changes that
pass around us in this world of changes, and
yet we cannot but feel sorry for the Cedar of
Lebanon.

Such is the history of the introduction of the
Cedar into France; a tale which has often been
told, but nowhere in a more pleasing manner
than in the foregoing extract from Sharpe's
"London Magazine." The date of its introduction into Great Britain is not known.

Spenser describes the tree, but in a way which
proves that he knew it only by name:

> " High on a hill a goodly Cedar grew,
>   Of wondrous length and straight proportion,
> That far abroad her dainty odours threw,
>   'Mongst all the daughters of proud Lebanon."

Milton's description is scarcely more accurate,
the Cedar being remarkable not on account of its
height, but the spread of its branches:

> " over head up grew,
> Insuperable height of loftiest shade,
> Cedar, and Pine, and Fir, and branching Palm."

The older botanists who mention it, speak of it as
a tree which they had never seen.

Evelyn, who wrote in the reign of Charles II.,
describes it as a beautiful and stately tree, clad
in perpetual verdure, and likely to thrive in Old
England as well as it did on the mountains of
Libanus, whence, he says, he had received cones
and seeds from the few remaining trees.   He

even mentions having frequently raised it from seeds, but whether these seeds came from Mount Lebanon, or the "Summer Islands" (the Bermudas), is not clear. It was undoubtedly his opinion that the Cedar of Lebanon and the Bermuda Cedar were identical; and though he mentions his having received cones and seeds from Palestine, he does not state that he raised any plants from these seeds. On the other hand, he says,—" I have frequently raised it from the seeds and berries, of which we have the very best in the world from the Summer Islands." In a letter written to the Royal Society, some twenty years after this, in which he describes the effects of a severe winter, he says,—" As for exotics, my Cedars, I think, are dead." Now it is not likely that Cedars of Lebanon, twenty years old, were killed by frost; but it is exceedingly probable that Bermuda or Pencil Cedars, which are very delicate, were destroyed, if inadequately protected.

I cannot, therefore, agree with Loudon that Evelyn introduced the Cedar, especially as there is no traditional existence of any tree planted by him. That it was introduced soon afterwards, and perhaps in consequence of his recommendation, there can be no doubt; for in 1683 four Cedars were planted in the Chelsea Garden, two of which are yet standing. These, and another tree which formerly stood at Farley, near Salisbury, are said by Lord Holland to have been introduced by his ancestor, Sir Stephen Fox. According to a tradition in the family of Ashby, of Quenby Hall in Leicestershire, a fine Cedar, which is yet standing in their grounds, was introduced between 1680 and 1690.

The Cedar first produced cones in England in the Chelsea Garden about 1766, since which time, vast numbers of trees have been raised both from native as well as foreign cones.

The Cedars of Lebanon are frequently mentioned in the Sacred Volume, and from their majestic growth are made an emblem of regal state, and so of the prosperity of the kingdom typified. They were formerly very abundant, but being much sought after for their timber, which was considered imperishable, their number is now greatly diminished. It was used in great quantities in the building of the Temple and Solomon's Palace at Jerusalem, and by the Tyrians the masts of ships were made of Cedar. The needle-shaped leaves are shorter than those of the Scotch Fir, and grow in bunches of more than twenty, like those of the Larch, but they are of a firmer texture, and are not deciduous. The cones, which stand erect, and in their young state are very conspicuous, are of a bright green colour and an oval shape; they adhere firmly to the branches, which are covered with a greyish brown bark. The horizontal branches, which are very large in proportion to the size of the trunk, are arranged in distinct layers or stages, and form a broadly pyramidal head. The extremities of the lower branches generally droop so as almost to touch the ground, when the tree stands alone; but if planted in masses, it bears a clean straight trunk, crowned by a depressed head. The beauty of the tree consists in the strength and elegant symmetry of its widely spreading branches. The resin which exudes from the stem and cones, is said to be as soft as balsam; the smell is very

similar to that of the balm of Mecca. Every
thing, indeed, about this tree has a strong
balsamic perfume, and hence, the whole forest
is so perfumed with fragrance that a walk

CONES OF CEDAR.

through it is delightful. This is probably the
" smell of Lebanon" to which reference is made
in Hosea xiv. 6.

So durable was Cedar-wood considered by the
ancients, that " to be worthy of being kept in
Cedar," *dignus cedro*, passed into a proverbial
expression for anything thought worthy of immortality. An oil extracted from it and called

cedreum, was said to render imperishable all
substances which were anointed with it.

The value of the timber of the Cedar, as
a building material, is now thought to have
been overrated by the ancients. It is reddish
white, with streaks, and does not seem to be
much harder than deal. It is sweet-scented
only for the first year after its being felled:
it soon begins to shrink and warp, and is said
to be by no means durable. But this is rather
the character of English-grown Cedar than of
timber which has come to maturity in its native
mountains.

" The prophet Ezekiel," Gilpin remarks, " has
given us the fullest description of the Cedar:
' Behold the Assyrian was a Cedar in Lebanon,
with fair branches and with a shadowing shroud,
and of an high stature; and his top was among
the thick boughs. His height was exalted above
all the trees of the field, and his boughs were
multiplied, and his branches became long because
of the multitude of waters when he shot forth.
Thus was he fair in his greatness, in the length
of his branches: the Fir-trees were not like his
boughs, and the Chestnut-trees [supposed to be
the Plane] were not like his branches.'*

" In this description, two of the principal characteristics of the Cedar are marked : the first is the
multiplicity and length of its branches. Few
trees divide so many fair branches from the
main stem, or spread over so large a compass
of ground. ' His boughs are multiplied,' as
Ezekiel says, ' and his branches became long,'
which David calls spreading abroad. His very

* Ezek. xxxi. 3—8.

boughs are equal to the stem of a Fir or a
Chestnut. The second characteristic is, what
Ezekiel, with great beauty and aptness, calls his
' shadowing shroud.' No tree in the forest is
more remarkable than the Cedar for its close-
woven, leafy canopy. Ezekiel's Cedar is marked
as a tree of full and perfect growth, from the
circumstance of its top being among the thick
boughs; that is, no distinction of any showy
head or leading branch appears; the head and
the branches are all mixed together. This is
generally in all trees the state in which they
are most perfect and most beautiful, and this
is the state of Ezekiel's Cedar. But though
Ezekiel has given us this accurate description of
the Cedar, he has left its strength, which is its
chief characteristic, untouched. But the reason
is evident; the Cedar is here introduced as an
emblem of Assyria, which, though vast and
wide-spreading, and come to full maturity, was,
in fact, on the eve of destruction. Strength,
therefore, was the last idea which the prophet
wished to suggest. Strength is a relative term,
compared with opposition. The Assyrian was
strong compared with the powers on earth; but
weak, compared with the arm of the Almighty,
which brought him to destruction. So his type,
the Cedar, was stronger than any of the trees
of the forest; but weak in comparison with
the axe which cut him off and left him (as
the prophet expresses the vastness of his ruin)
spread upon the mountains and in the valleys,
while the nations shook at the sound of his fall.
Such is the grandeur and form of the Cedar of
Lebanon."

Southey alludes in the following lines to a singular superstitious belief entertained by the Maronites of Mount Lebanon:

> " It was a Cedar-tree
> Which woke him from that deadly drowsiness;
> Its broad, round-spreading branches, when they felt
> The snow, rose upward in a point to heaven,
> And standing in their strength erect,
> Defied the baffled storm."   THALABA.

" The Maronites say that the snows have no sooner begun to fall, than these Cedars, whose boughs, in their infinite number, are all so equal in height that they appear to have been shorn, and form, as it were, a sort of wheel or parasol— than these Cedars, I say, never fail at that time to change their figure. The branches, which before spread themselves, rise insensibly, gathering together, it may be said, and turn their points upwards towards Heaven, forming altogether a pyramid. It is Nature, they say, who inspires this movement, and makes them assume a new shape, without which these trees never could sustain the immense weight of snow remaining for so long a time."*

The Cedar is a native not only of the mountain from which it derives its name, but of Northern Africa, where it was found in abundance by Mr. Drummond Hay. Of the many accounts that have been published of the famous grove of Cedars on Mount Lebanon, it will be sufficient to quote the following:—" These noble trees grow amongst the snow, near the highest part of Libanus; and are remarkable as well for their own age and largeness, as for the frequent allusions made to them in the Word of God.

* De la Roque, 1772.

Here are some very old, and of a prodigious bulk; and others younger, of a smaller size. Of the former I could reckon up only sixteen; the latter are very numerous. I measured one of the largest, and found it twelve yards six inches in girth, and yet sound, and thirty-seven yards in the spread of its boughs. At about five or six yards from the ground, it was divided into five limbs, each of which was equal to a great tree."*

" We are informed, from the ' Memoirs of the Missionaries in the Levant,' that, upon the day of Transfiguration the Patriarch of the Maronites (Christians inhabiting Mount Libanus), attended by a number of bishops, priests and monks, and followed by five or six thousand of the religious from all parts, repairs to these Cedars, and there celebrates the festival that is called ' The feast of Cedars.' We are also told, that the Patriarch officiates pontifically on this solemn occasion; that his followers are particularly mindful of the Blessed Virgin on this day, because the Scripture compares her to the Cedars of Lebanon; and that the same Holy Father threatens with ecclesiastical censure those who presume to hurt or diminish the Cedars still remaining."†

" The famous Cedars of Lebanon are situated on a small eminence, in a valley at the foot of the highest part of the mountain. The land on the mountain's side has a sterile aspect, and the trees are more remarkable, as they stand altogether in one clump, and are the only trees to be seen in this part of Lebanon. There may be about fifty of them, but their present appear-

* Maundrell.　　　　　　　　† Dr. Hunter.

ance ill corresponds with the character given of them in Scripture. There was not one of them at all remarkable for its dimensions or beauty; the largest among them is formed by the junction of four or five trunks into one tree. Numerous names carved on the trunk of the larger trees, some with dates as far back as 1640, record the visits of individuals to this interesting spot, which is nearly surrounded by the barren chain of Lebanon, in the form of an amphitheatre of about thirty miles circuit, the opening being towards the sea."*

" These trees are the most renowned natural monuments in the universe; religion, poetry, and history, have all equally celebrated them. The Arabs of all sects entertain a traditional veneration for these trees. They attribute to them, not only a vegetative power, which enables them to live eternally, but also an intelligence, which causes them to manifest signs of wisdom and foresight, similar to those of instinct and reason in man. They are said to understand the changes of seasons; they stir their vast branches as if they were limbs; they spread out or contract their boughs, inclining them towards heaven or towards earth, according as the snow prepares to fall or melt. These trees diminish in every succeeding age. Travellers formerly counted thirty or forty; more recently, seventeen; more recently still, only twelve. There are now but seven.† These, however, from their size and general appearance, may be fairly presumed to have existed in biblical

* Irby and Mangles.
† Warburton maintains that there are still twelve of the very largest trees, and about a thousand of all ages.

times. Around these ancient witnesses of ages long since past, there still remains a little grove of yellower Cedars, appearing to me to form a group of from four to five hundred trees or shrubs. Every year, in the month of June, the inhabitants of the neighbouring valleys and villages climb up to these Cedars and celebrate mass at their feet. How many prayers have resounded under these branches! and what more beautiful canopy for worship can exist?" *

* Lamartine.

DEODAR AT ENGLEFIELD, BERKS.

THE DEODAR.

CEDRUS DEODARA.

THE Deodar, Holy Cedar, or Himalayan Cedar, is known to us only as an ornamental plant of exquisitely beautiful outline and graceful spray, giving an air of refinement to every lawn and shrubbery to which it has been admitted; but in its native haunts it is a magnificent tree, of rapid growth and enormous size, with the evergreen beauty of the Cedar of Lebanon when living, and affording when cut down, timber not simply durable, but imperishable. No wonder, then, that the untaught Hindoos should look on it with reverence, giving it a name expressive of this feeling, "the gift of God," and in some districts using its fragrant wood as a material for their temples, and burning it as incense on occasions of great ceremony.

The leaves and cones are very like those of the Cedar of Lebanon; but the general habit of the two trees is different in every stage of their growth. When young, the Deodar resembles a luxuriant Larch with a leafy base, but the branches are more delicate and thickly clothed with foliage, and the extremities of all the shoots, even the leader, droop most gracefully. What will be the appearance of the full-grown tree in this climate it is impossible to conjecture. If it succeeds, which it gives every prospect of doing, it will prove one of the most valuable additions that has ever been made to the trees of Britain, both for the sake of its picturesque beauty and its timber.

In its native state, the Deodar grows high up on the slopes of the Himalayan chain, attaining an enormous size and hanging the sides of the mountains with a perennial coat of verdure. It is not unusual to see it in favourable situations with a girth varying from twenty-four to thirty feet, with a proportionate height and vast expanse. No adequate notion can be formed of the majestic character of the tree from the small-sized specimens now in existence in England. The Deodar varies in appearance greatly during its growth. The young tree looks a good deal like the Larch, rising in an elongated conical mass, tapering off into a bold leading shoot. When it attains a height of fifty or sixty feet, the terminal leader withers, the top becomes flattened, the lateral growth is increased, and the tree drops the character of the Larch, and puts on that of the Cedar. So much does its appearance alter, that the English residents at the hill stations, like Simla, imagine that there are two species: the old tree they call the Deodar, and the younger one, the Kelon.* Nothing can exceed the grandeur of an old Deodar of thirty feet girth. The branches begin to spread horizontally close to the ground, rising flight above flight in successive sheeted steps into a rounded or slightly flattened top. Seldom or never is the slightest trace of decay seen in the trunk, and the tree never, except when

* Maddon says that the Kelon of Simla, is the true Deodar, and that the sacred name is there given to a species of Cypress, *Cupressus torulosa*.

growing in very exposed situations, puts on the depressed, abrupt character of the Cedar of Lebanon, Others of the Fir tribe may compete with it in height and dimensions, but in economic value it, beyond all question, occupies the first place. The wood is light, strong and compact, straight in the grain, free from knots, easily wrought, and highly perfumed with a most delightful aromatic perfume, which it never loses. In durability it is certainly without a rival; Kyanized by the hand of nature, it defies wind and weather, resisting the soaking rains of the Himalayan mountains for ages. Rot, under any aspect, is unknown to it. You will see in the Himalayas Deodar timbers built into the walls of old temples now levelled nearly to their foundations; the surface bleached and ragged, but the body of the wood undecayed, and emitting its characteristic odour fresh as ever. In Cashmeer the pillars which support the roof of the great Mosque, built in the days of our later Henries, are formed of Deodar trees stripped of their bark; they exhibit not a crack or sign of decay, and still smell like pencil-wood. All the boats in the valley are built of Deodar, and when they get crazy at the joinings by age, the old planks have their weathered surface planed off by the adze, and are then undistinguishable from the new wood, along with which they are rebuilt. The wood is so straight and equal in the grain, that it gives planks three feet broad simply by the action of the wedge. Timber-saws are unknown in the Himalayas. It is hardly possible to overrate its value as a timber-tree, or the advantages that would follow from getting it established in Britain, where there is every prospect of its doing well. It grows fast in favourable situations, sometimes making shoots two feet long in a single season.

Bishop Heber, in a letter to Lord Grenville, giving an account of a visit which he paid to the Himalayan Mountains, describes it as a "splendid tree, with gigantic arms and dark narrow leaves, which is accounted sacred, and is chiefly seen in the neighbourhood of ancient Hindoo temples, and which struck my unscientific eye as nearly resembling the Cedar of Lebanon.* I found it flourishing at nearly nine thousand feet above the level of the sea, and when the frost was as severe at night as is usually met with at the same season in England."

Mr. Moorcroft gives the following proofs of the durability of the timber: "A few years ago a building, erected by the order of the Emperor Akbar, probably about 1597, was taken down, and its timber, which was that of the Deodar, was found so little impaired as to be fit to be employed in a house built by Rajah Shah. Its age must then have been two hundred and twenty-five years." He also describes a mausoleum, which was erected nearly four hundred years since, the walls of which are of brick and mortar, strengthened with beams of Deodar. In this last instance, the sap-wood, which had been carelessly left in some places, had been pierced by a worm to the depth

---

* So closely do the Cedar and Deodar resemble each other in botanical characters, that Sir W. J. Hooke says: "No botanist, in describing the trees, has given clear and distinctive botanical characters." Captain Munro, who has travelled much among the Deodars of the Himalaya, considers them mere varieties of the Cedar of Lebanon.

of a quarter of an inch, but the heart-wood, not-withstanding this long exposure to the weather, was neither crumbly nor worm-eaten, the only perceptible effect being, that the surface was jagged, from the softer parts of the wood having been often washed by the rain. He also obtained specimens of the wood from a bridge in Ladakh, which had been exposed to the water for nearly four hundred years. It has a remarkably fine close grain, capable of receiving a very high polish ; so much so indeed, that a table formed of the section of a trunk nearly four feet in diameter, has been compared to a slab of brown agate.

It is readily propagated by seeds, and may be raised also from cuttings. It has also been grafted on the Larch, but can scarcely be expected to attain perfection while dependent on the roots of a deciduous tree, the duration of which, compared to its own, is very limited. It has also been grafted on the Cedar of Lebanon, with a better chance of success. The country is indebted for the first introduction of Deodar seeds to the Hon. William Leslie Melville, who brought home some cones in 1831, and supplied seeds to the Horticultural Society, &c. By the liberality of the East India Company they have since been imported in large quantities, and trees are now so abundant, that although they were sold in 1838 at the high price of two guineas each, seedlings two years old may now be purchased at the rate of four shillings a dozen.

The largest plantation of Deodars which has been made in Europe is that of W. Ogelvie, Esq., secretary of the Zoological Society, who, on his estate of Altinachree in Tyrone, has planted eleven acres.

ARAUCARIA.

THE CHILI PINE.

ARAUCARIA IMBRICATA.

WHEN the numerous trees, which have been introduced into Great Britain during the last fifty or sixty years, have attained to perfection and in a measure altered the features of artificial landscape scenery, no tree will contribute so much to produce this effect as the Araucaria or Chili Pine. Seen from a distance, it has just enough of the character of the Fir tribe to point out its relationship, but is unlike them all; and when inspected more closely, it bears not the least resemblance to any tree known in Britain, and, even to any eye but that of a botanist, it no longer has any affinity with the Firs.

In its native haunts, the Cordillera chain in Chili, it is a lofty tree, exceeding a hundred feet in height, with a straight trunk, covered with a thick cork-like bark, which abounds with resin. The branches are longest and most numerous near the base; thus the tree has a pyramidal form. The leaves are broad, rigid, tough, and sharp pointed, remaining attached to the tree for many years. The branches are cylindrical and thickly covered by the clasping leaves, resembling, as Loudon happily remarks, " in young trees, snakes partly coiled round the trunk, and stretching forth their long slender bodies in quest of prey." The Araucaria, as we are acquainted with it in England, is rather a singular than a beautiful tree, giving a foreign air to every place where it is planted, but not possessing elegance of form. On the steep rocky ridges of the South American

BRANCH OF CHILI PINE.

mountains, it would seem to be no less majestic than singular. "When we arrived at the first Araucarias the sun had just set; still some time remained for their examination. What first struck our attention, were the thick roots of these trees, which lie spread over the stony and nearly naked soil like gigantic serpents, two or three feet in thickness; they are clothed with a rough bark, similar to that which invests the lofty pillar-like trunks, of from fifty to a hundred feet in height. The crown of foliage occupies only about the upper quarter of the stem, and resembles a large depressed cone. The lower branches, eight or twelve in number, form a circle round the trunk; they diminish till they are but four or six in a ring, and are of most regular formation, all spreading out horizontally and bending upwards only at their tops. They are thickly invested with leaves that cover them like scales, and are sharp-pointed, above an inch broad, and of such a hard and woody texture that it requires a sharp knife to sever them from the parent stock. The general aspect of the Araucaria is most striking and peculiar, though it undoubtedly bears a distant family likeness to the Pines of our country. The fruit placed at the ends of the boughs, are of regular globular form, as large as a man's head, and each consists of beautifully imbricated scales, that cover the seeds, which are the most important part of this truly noble tree. Such is the extent of the Araucarian forests, and the amazing quantity of nutritious seeds that each full-grown tree produces, that the Indians are ever secure from want: and even the discord that prevails frequently among the different hordes, does not prevent the quiet collection of this kind of harvest. A single fruit contains between two and three hundred kernels; and there are frequently twenty or thirty fruits on one stem; and, as even a hearty eater among the Indians except he should be wholly deprived of every other kind of food cannot consume more than two hundred nuts in a day, it is obvious that eighteen Araucarias will maintain a single person for a whole year. The kernel,

CONE OF CIHLI PINE.

which is of the shape of an almond, but double the size, is surrounded with a tough membrane, which is easily removed; though relishing when prepared, it is not easy of digestion, and containing but a small quantity of oil it is apt to cause disorders in the stomach with those who are not accustomed to this diet. The Indians eat them either fresh, boiled, or roasted; and the latter mode of cooking gives them a flavour something like that of a chestnut. For winter's use, they are dried after being boiled; and the women prepare a kind of flour and pastry from them. The collecting of these fruits would be attended with great labour, if it were always necessary to climb the gigantic trunks, but as soon as the kernels are ripe, towards the end of March, the cones drop off of themselves, and, shedding their contents on the ground, scatter liberally a boon which nothing but the little parrot and a species of cherry-finch divide with the Indians. In the vast forests, of a day's journey in extent, that are formed by these trees, in some districts the fruits lie in such plenty on the ground, that but a small part of them can be consumed.

The wood of the Araucaria is white, and towards the centre of the stem bright yellow. It yields to none in hardness and solidity, and might prove valuable for many uses if the places of growth of the tree were more accessible. It has been used by the Spaniards for ship-building; but it is much too heavy for masts. If a branch be scratched, or the scales of an unripe fruit be broken, a fragrant milky juice immediately exudes, that soon changes to a yellowish resin, which is considered by the Chilians as possessing such medicinal virtues, that it cures the most violent rheumatic headaches when applied to the spot where the pain is felt.

The Araucaria was first introduced into England by Menzies, who accompanied Vancouver in his expedition to Chili in 1795. Living specimens were given by him to Sir Joseph Banks, one of which is still growing at Kew. It was at first supposed to be delicate, and was protected from frost during winter, but it suffered from this mode of treatment, and having been deprived of its lower branches, the character of the tree was destroyed. One of the finest and handsomest trees in England stands in the garden of R. Dawson, Esq., Tottenham. It is twenty feet high; the branches descend to the ground on all sides, and the main stem is densely covered with leaves down to the very roots. It was planted out at the height of four inches in April, 1832.

A writer in the " Gardener's Chronicle" recommends that the seeds of this tree, when
planted, should not be buried beneath the ground,
but simply laid on the surface, with a small
quantity of earth raised around them, but not so
as to cover them.

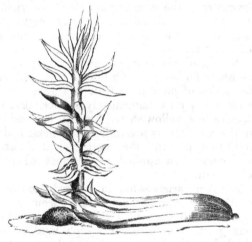

SEEDLING ARAUCARIA.

Young plants, which in 1838 were sold at from
three to five guineas each, may now be purchased
for about two shillings each.

Although the Araucaria, of which we have been
speaking, derives its name from the Araucanos, a
people of Chili, and therefore, strictly speaking,
is a South American tree, there are two other
species which inhabit the Eastern world.   One of
these, the Norfolk Island Pine, *Araucaria excelsa*,
ranks among the most magnificent of known trees,

rising to the almost incredible height of two hundred and twenty feet, with a trunk thirty feet in circumference. Mrs. Meredith, in her entertaining "Notes and Sketches of New South Wales," thus describes some young trees which grew near her house in that country:—" The Norfolk Island Pine is certainly the most noble and stately tree of all the Pine family that I have ever seen, beautiful as they all are. The tall, erect and tapering stem, the regularity of the circling branches, lessening by small degrees from the widely-spread expanse below, to the tiny cross that crowns the summit of the exquisite natural spire, and the really verdant, dense, massive foliage, clothing the whole with an unfading array of scale armour, form altogether the finest model of a Pine that can be imagined. The cones too are worthy to grow on such a tree; solid ponderous things, as large as a child's head—not a baby's head neither—with a fine embossed coat of mail, firmly seated on the beam-like branches, as if defying the winds to shake them. Mr. Meredith climbed very nearly to the summit of our tallest Pine, and said he had never seen anything more beautiful than the downward view into and over the mass of diverging branches spread forth beneath him. He brought me down one cone with its spray, if I may so call the armful of thick green shoots that surrounded it, and I was gazing on it for half the day after; it was so different from anything I had ever seen before, so new, and so grandly beautiful. The rigidity of the foliage had a sculpture-like character that made me think how exquisitely Gibbons would have wrought its image in some of his graceful and stately designs, had he ever seen the glorious tree. One grew near to the front verandah, and some of its enormous roots had spread under the heavy stone pavement, lifting it up in an arch, like a bridge. When the cones ripened, the large winged seeds fell out in great numbers; they require to be planted immediately, as the oil in them quickly dries up, and with it the vegetative properties are lost."

It was introduced into England in 1793, but as it requires protection during the winter, its dimensions must be limited to the size of the conservatory in which it stands. " It is a highly interesting fact," says Dr. Lindley, " that a plant very nearly the same as this Araucaria certainly once grew in Great Britain. Remains of it have been found in the lias of Dorsetshire, and have been figured under the name of *Araucaria primæva.*

THE JUNIPER.

JUNIPERUS COMMUNIS.

Class—DIŒCIA.   *Order*—MONADELPHIA.

THE Juniper is well known to the readers of the
English version of the Bible as the tree under
which the prophet Elijah, wearied with his journey
through the wilderness, sat down to rest, when
flying from the persecution of Jezebel (1 Kings
xix. 4). This tree, or rather shrub, is generally
supposed to be a species of Broom (*Genista
monosperma*) which is one of the few plants to be
found in the Arabian deserts. Burckhardt mentions
it as growing also in the deserts to the south of
Palestine, so that the Juniper which sheltered the
prophet may possibly be the tree in question,
though other travellers have looked for it in the
neighbourhood of Mount Horeb, instead of at the
distance of a day's journey from Jerusalem. Lord
Lindsay speaks of his having frequently sheltered
himself under a Broom in the valleys of Mount
Sinai, an incident which Dr. Kitto fixes on as
conclusive, seemingly forgetting that Elijah was
as yet distant a journey of forty days from the
same spot. Nevertheless the similarity of the
Hebrew name Rothem to the Arabic Rethem,
makes it highly probable that the two trees are
identical.

II.   F F

The Juniper is a native of all the northern parts
of Europe, and in Great Britain is generally found
on hills and heathy downs especially where the

THE JUNIPER.

soil is chalky. It usually appears as a bushy, evergreen shrub, with narrow sharp-pointed leaves,
which are arranged in threes, and are of a glaucous hue above, and dark green beneath. Instead

of bearing dry cones, like most other trees belonging to this tribe, the Juniper produces fleshy
berries which are formed of the united scales of the
calyx, and contain three oblong seeds. The barren
flowers are small, and grow on separate plants from
the fertile flowers, in the axils of the leaves; the
Juniper was consequently placed by Linnæus in a
different class from the rest of the Fir tribe, though
naturally closely allied to them.

The Juniper of the ancients was probably a
different species from that which is indigenous to
Britain. The common species, in Evelyn's time,
was frequently transplanted from the open commons to make hedges and arbours. The berries
were used as a spice, and were also employed
medicinally. " If it arrive to full growth, spits
and spoons, imparting a grateful relish, and very
wholesome where they are used, are made of this
wood, being well dried and seasoned: and the very
chips render a wholesome perfume within doors,
as do the dusty blossoms in spring, without."
Phillips says, that on the Continent both the wood
and berries are burnt to fumigate the rooms of the
sick. In Sweden the berries are made into a con-
serve and eaten at breakfast. In some places they
are roasted and used as a substitute for coffee.
The heathcock of Germany, he also says, is not
eatable in autumn, being so strongly flavoured with
Juniper berries, on which it then feeds. The
principal use of the berries at present, is to flavour
hollands or geneva, a spirit distilled from corn.
In the manufacture of London gin (a corruption
of geneva) oil of turpentine is said to be substituted
for Juniper berries, and is perhaps one of the least
noxious ingredients.

The wood will take a high polish, but is rarely
to be obtained of a sufficient size for useful purposes. Loudon mentions some trees which have
attained a height of from sixteen to thirty feet.

The Pencil-Cedar, *Juniperus Bermudiana,* is a
native of the island from which it derives its specific
name.

RUIN OF AN OLD OAK IN ALDERMASTON PARK, BERKS.

INDEX.

THE END.

LONDON :

Printed by S. & J. BENTLEY and HENRY FLEY,
Bangor House, Shoe Lane.

Printed in the United States
By Bookmasters